# Lecture Notes in Mobility

**Series editor**

Gereon Meyer, Berlin, Germany

More information about this series at http://www.springer.com/series/11573

Dirk Fornahl · Michael Hülsmann
Editors

# Markets and Policy Measures in the Evolution of Electric Mobility

Springer

*Editors*
Dirk Fornahl
University of Bremen
Bremen
Germany

Michael Hülsmann
Jacobs University Bremen
Bremen
Germany

ISSN 2196-5544 ISSN 2196-5552 (electronic)
Lecture Notes in Mobility
ISBN 978-3-319-24227-9 ISBN 978-3-319-24229-3 (eBook)
DOI 10.1007/978-3-319-24229-3

Library of Congress Control Number: 2015950059

Springer Cham Heidelberg New York Dordrecht London


Printed on acid-free paper

Springer International Publishing AG Switzerland is part of Springer Science+Business Media
(www.springer.com)

# Acknowledgments

This book is based on the conference "Future mobility. Markets and policy measures in the evolution of electric mobility" which was arranged by the Centre for Regional and Innovation Economics at OFFIS e.V. in Oldenburg in cooperation with Jacobs University in December 2013. The conference and the associated project was funded by the Federal Ministry of Transport and Digital Infrastructure (BMVI) in the context of the "Model Regions Electric Mobility". Responsible for programme coordination was the NOW GmbH National Organisation Hydrogen and Fuel Cell Technology.

This book is, in a sense, the successor of the edited volumes "Evolutionary Paths Towards the Mobility Patterns of the Future" (Hülsmann and Fornahl 2014) which was based on the conference held in the same context in the year 2011 and "Future mobility. Theoretical, empirical and political aspects of the first stage of electric mobility evolution" (Hülsmann and Fornahl, forthcoming) which was based on the conference held in the same context in the year 2012.

Therefore, we are indebted to the Ministry of Transport and Digital Infrastructure (BMVI) that provided us with the opportunity to organize the conference and brought scientists working on electric mobility together to discuss ideas and to set up this book. Furthermore, we want to thank all presenters and participants of the conference for their interesting presentations and fruitful discussions.

# Contents

**Introduction** . . . . . . . . . . . . . . . . . . . . . . . . . . . . . . . . . . . 1
Antje Campen, Dirk Fornahl, Nikolai Iliew and Michael Hülsmann

**Part I Client's Perspectives: Threats and Opportunities**

**Strategic Perspectives of Electric Mobility—Steps Towards the Slope of Enlightenment** . . . . . . . . . . . . . . . . . . . . . . . . . . . . 9
Richard Colmorn and Michael Hülsmann

**On the Road to an Electric Mobility Mass Market—How Can Early Adopters be Characterized?** . . . . . . . . . . . . . . . . . . . . . . . . . . . . 21
Axel Ensslen, Alexandra-Gwyn Paetz, Sonja Babrowski, Patrick Jochem and Wolf Fichtner

**How to Increase the Attractiveness of Electric Cars as Fleet Cars?** . . . 53
Jasmin Nehls

**Flexible Carsharing—Potential for the Diffusion of Electric Mobility** . . . . . . . . . . . . . . . . . . . . . . . . . . . . . . 67
Sandra Wappelhorst, Julia Dobrzinski, Andreas Graff, Josephine Steiner and Daniel Hinkeldein

**Part II Policy Perspectives**

**Public E-Mobility Funding and its Influence on Cooperation and Innovation** . . . . . . . . . . . . . . . . . . . . . . . . . . . . . . 87
Isa Reinecke

**China's Leapfrog to New Electric Vehicles** . . . . . . . . . . . . . . . . . . 103
Peter Thomas in der Heiden

**Part III Management Perspectives**

**Organizations as Change Agents Towards New Modes of (Sustainable) Mobility: Insights from Practice Oriented Projects** . . . 131
Dirk Dalichau and Birgit Blättel-Mink

**Competition, Cooperation, or Both? Designing Coopetitive Structures in the German Automotive Supply Industry** . . . . . . . . . . . . . . . . . . 145
Lukas Schilling, Thomas M. Fojcik and Konrad Wirmer

**Valuing Battery Degradation from a Battery Lessor's Perspective** . . . . 163
Björn Hildebrandt, Sebastian Busse, Simon Trang and Lutz M. Kolbe

**The Influence of Organisational and Structural Parameters of Commercial Fleets on the Substitution Potential of Battery Electric Vehicles** . . . . . . . . . . . . . . . . . . . . . . . . . . . . . . 179
Heike Flämig, Christian Matt, Christian Rudolph and Sören Christian Trümper

**Who Wants to Be a Millionaire? The Lead Market Potential of Electric Mobility** . . . . . . . . . . . . . . . . . . . . . . . . . . . . . 195
Noreen E. Werner

# Introduction

**Antje Campen, Dirk Fornahl, Nikolai Iliew and Michael Hülsmann**

**Abstract** For many years electric mobility is discussed as a possible option to cope with problems related to urbanization, greenhouse gas emissions, especially carbon dioxide and as a possible solution to reduce the dependencies on fuels from fossil sources. The generation of electricity from renewable resources reduces greenhouse gas emissions considerably, which means that driving an electric vehicle on renewable energy might bring down emissions close to zero. This chapter introduces electric mobility issues and provides an overview of the contributions of this book.

## 1 Introduction

For decades electric mobility is discussed as a solution for future mobility. Many hopes are pinned on new technologies that are supposed to provide more sustainable forms of transportation.

The dependence on fossil fuels and the climate change led not only in Germany to a debate on how the future of mobility might be designed. In terms of an increasing world population, further worldwide urbanization, economic growth as well as climate change on the whole it is common sense that in a few decades the oil era comes to an end. Combustion engines, based on fossil fuels can no longer

A. Campen (✉) · D. Fornahl · N. Iliew
Centre for Regional and Innovation Economics, Universität Bremen, Wilhelm-Herbst-Str. 12, 28359 Bremen, Germany
e-mail: acampen@uni-bremen.de

D. Fornahl
e-mail: dirk.fornahl@uni-bremen.de

N. Iliew
e-mail: n.iliew@uni-bremen.de

M. Hülsmann
Jacobs University Bremen, Campus-Ring 1, 28759 Bremen, Germany
e-mail: m.huelsmann@jacobs-university.de



D. Fornahl and M. Hülsmann (eds.), *Markets and Policy Measures in the Evolution of Electric Mobility*, Lecture Notes in Mobility,
DOI 10.1007/978-3-319-24229-3_1

satisfy all the future mobility needs. As one option Electric mobility emerges as a solution approach for individual transportation that might meet environmental and resource requirements as well as economic and social aspects.

The usage of electric vehicles—in combination with the use of renewable energies—can help to reduce emissions and the dependency on fossil fuels. Electric mobility is also a driver of innovation and the establishment of new technologies, which in turn help to create and maintain jobs and generate prosperity (Pehnt et al. 2007; Schindler et al. 2009; Whitmarsh and Koehler 2010).

Therefore, electromobility does not only constitute a field of scientific interest, but is strongly promoted by policy makers (Bundesregierung 2011). The changed conditions lead the automobile manufacturers, customers and legislature to realize that in the future there are more options as only Internal Combustion Engines (ICE). The emerging path is still in the beginning, meaning there are lots of alternatives.

One of the potential solutions, full or hybrid electric vehicles, have to take into account the technical, economic and organizational requirements of customers, as well as how monetary and non-monetary support schemes could or should be designed to facilitate market introduction and diffusion. In this regard the issue of how strategies, structures and systems need to change in order to bring electric mobility to the market and how customer demand for a payable and sustainable mobility can be satisfied have to be taken into account, too.

## 2 Background and Structure

The contributions in this book are essentially based on presentations held at the workshop "Future mobility. Markets and policy measures in the evolution of electric mobility" in December 2013 which was arranged by the Centre for Regional and Innovation Economics (University of Bremen) at OFFIS e.V. in Oldenburg in cooperation with Jacobs University.[1] At that time, a large part of field trials had been completed and assessments about the direction of future developments of e-mobility could be made.

The focus of this volume is to give an overview on new insights and highlight trends and challenges for markets and policies in the field of electric mobility. The intention is to gather theoretical, empirical and political contributions from different fields, explicitly focussing on the market introduction phase of electric mobility in order to shed some more light on the complex demand, supply and policy side topics.

This book is subdivided into three parts. Part 1 discusses threats and opportunities of electric mobility from a client's perspective. Part 2 deals with policy

[1]This book is, in a sense, the successor of the edited volumes "Evolutionary Paths Towards the Mobility Patterns of the Future" (Hülsmann and Fornahl 2014) which was based on the conference held in the same context in the year 2011 and "Future mobility. Theoretical, empirical and political aspects of the first stage of electric mobility evolution" (Hülsmann and Fornahl, forthcoming) which was based on the conference held in the same context in the year 2012.

perspectives and how these facilitate market entry and diffusion. Finally part 3 shows the management and market perspectives and how strategies, structures and systems must change in order to bring electric mobility to the market.

## 3 Client's Perspectives: Threats and Opportunities

Part 1 of the book deals with the client's perspectives. The authors present the technical, economic and organizational requirements of customers and their acceptance and experiences regarding electric mobility. They want to shed light to the question if there is a strong willingness-to-pay for electric mobility.

**Ensslen et al.** direct their attention towards different field trials and corresponding acceptance studies with new technologies which have been carried out between 2010 and 2013 at the Chair of Energy Economics at the Karlsruhe Institute of Technology (KIT). Additionally, regression methods have been applied in order to characterize early involved Electric Vehicle (EV) adopters based on a subsample of EV company car users in the French-German context. A binary logit model explaining private EV purchase intention has been developed.

**Nehls** explores ways to increase the attractiveness of electric cars. The paper first analyzes promising applications of electric cars within company fleets, which are then aggregated to five key purchase motives. It then elaborates service concepts tackling image and Total Cost of Ownership (TCO) issues of the electric car. Discussion of added value reveals potential for the successful integration of electric cars into corporate fleets and includes several best practice examples.

**Wappelhorst et al.** present a study which was carried out in the city of Berlin comprising several surveys dealing with carsharing. Expectations and experiences of users with a flexible carsharing offer including exclusively electric vehicles were evaluated and analyzed over a period of 1 year.

**Colmorn and Hülsmann** deal with a realistic estimation of threats and opportunities regarding the usage of electric mobility. These issues can be identified with the help of the hype cycle. Here the authors rise the question about the steps which lead towards the "slope of enlightenment". As a result, the paper contributes to the research on future concepts of mobility by providing an overview of different strategic perspectives.

## 4 Policy Perspectives

Part 2 shows how policy measures can or shall support market introduction. The articles discuss how the efficiency of the funding can be increased and how monetary and non-monetary support schemes shall be designed to facilitate market entry and diffusion. In the following the contents of the articles are illustrated.

**Reinecke** deals with public e-mobility funding and its influence on cooperation and innovation. In a qualitative survey with different organizations involved in the innovation system of e-mobility, impediments as well as proposals for improvement were identified. The study focused on cooperative behavior, and it could be shown that political measures have a great impact on that.

**In der Heiden** analyzes a leapfrogging strategy from the Chinese government to jump-start technological upgrading in the country's automobile industry. The paper identifies this leapfrogging attempt as a strategic project initiated by the central government and organized in a top down manner. It sheds light on the supporting public policy framework and conducts a preliminary review of the strategy's accomplishments until mid-2012. By way of an indepth examination of guidance policies and incentive structures, the paper highlights the challenge of coordinating promotion plans across the country.

## 5 Management Perspectives

Part 3 of the book deals with the questions of whether electric mobility can really gain a considerable market share in the upcoming years and how strategies, structures and systems must change in order to bring electric mobility to the market. The customer demands for a payable and sustainable mobility must be satisfied. The following papers provide insights into research focussing on the role of the markets in the evolution of electric mobility.

**Dalichau and Blättel-Mink** concentrate on electric cars, on carsharing, and on intermodal services as possible ways of reducing emissions, decreasing the number of cars and overcoming the problem of insufficient space without at the same time increasing social inequality and threatening economic competitiveness. In an ongoing research project, the role of organizations as change agents towards sustainable mobility is being explored. The preliminary findings allow for a deeper understanding of how organizations understand their roles in a transformation process towards new modes of (sustainable) mobility, but they also indicate some of the problems that still exist.

**Fojcik et al.** investigate the process and the effectiveness of various organizational measures to balance opportunistic behavior in coopetitive situations, i.e. the simultaneity of cooperation and competition between at least two companies, in the context of the automotive supply industry. In a first step, various organizational measures of transaction cost, principal-agent and social exchange theory were extracted. In a second step, ten German automotive suppliers have been examined during coopetitive situations by means of a case study analysis in order to test the process, the efficiency and the effectiveness of the theoretically derived measures.

**Hildebrandt et al.** examine the degradation in lithium-ion batteries from a lessor's perspective. They suppose that battery life can be influenced directly by the user. In the case of a lease contract, the battery always remains the property of the lessor. In this respect, there are no direct incentives for the user to increase battery life by expending additional effort. Therefore, the idea of a dynamic leasing concept as an opportunity for car manufacturers to encourage consumers to adopt a gentle treatment of the battery in order to extend battery lifetime and thus the residual value is discussed.

**Trümper et al.** analyze commercial fleets in the City of Hamburg to investigate the influence of organizational parameters, such as vehicle use, as well as structural parameters, such as vehicle types and fleet size, on the potential of battery electric vehicles to substitute conventional ones.

**Werner** examines the conditions of global dominant innovation design considering e-mobility and the emergence of the lead market in different countries. With the help of a principal component analysis, different country-specific indicators were applied in the model of five lead market factors. But lead markets change quickly and the given theoretical concept is only estimating opportunities and not forecasting technology developments.

**Acknowledgements** This article was supported with funds from the Federal Ministry of Transport and Digital Infrastructure (BMVI). Responsible for programme coordination was the NOW GmbH National Organisation Hydrogen and Fuel Cell Technology.

## References

Bundesregierung (2011) Regierungsprogramm Elektromobilität. Accessed http://www.bmbf.de/pubRD/programm_elektromobilitaet.pdf

Hülsmann M, Fornahl D (eds) (2014) Evolutionary paths towards the mobility patterns of the future. Springer, Berlin

Hülsmann M, Fornahl D (eds) (forthcoming) Future mobility. Theoretical, empirical and political aspects of the first stage of electric mobility evolution. Springer, Berlin

Pehnt M, Höpfner U, Merten F (2007) Elektromobilität und erneuerbare Energien. Arbeitspapier Nr. 5. Accessed http://www.bmu.de/files/pdfs/allgemein/application/pdf/elektromobilitaet_ee_arbeitspapier.pdf

Schindler J, Held M, Wuerdemann G (2009) Postfossile Mobilität – Wegweiser für die Zeit nach dem Peak Oil. VAS- Verlag für Sozialwissenschaften, Bad Homburg

Whitmarsh L, Koehler J (2010) Climate change and cars in the EU: the role of auto firms, consumers, and policy in responding to global environmental change. Cambridge J Reg Econ Soc 3(3):427–441

### Author Biographies

**Antje Campen** is research associate at the Centre for Regional and Innovation Economics (CRIE) at the University of Bremen. Her work is centered on cluster analysis, innovation processes and regional strategies. She holds a graduate degree in Geography.

**Dirk Fornahl** leads the Centre for Regional and Innovation Economics (CRIE) at the University of Bremen. His work is settled on cluster analysis, analysis of regional supply chains and potential, innovation processes and networks. He also coordinates the socio-economic research in the electric model region Bremen/Oldenburg.

**Nikolai Iliew** is research assistant at the Centre for Regional and Innovation Economics (CRIE) at the University of Bremen. He is a student of Business Administration at the University of Bremen with focus on Supply Chain Management, Network Design and Operations research.

**Michael Hülsmann** is Managing Director at Jacobs University Bremen gGmbH since June 2014. From 2009 until 2014 he led the department "System Management" at Jacobs University. His main research focus lies on a Strategic Management of Logistics Systems. Thereby, Prof. Hülsmann is doing research mainly in the fields of competence and technology based positioning, accomplishing of technological change in value adding networks and setting up of inter-organizational coordination mechanisms.

# Part I
# Client's Perspectives: Threats and Opportunities

# Strategic Perspectives of Electric Mobility—Steps Towards the Slope of Enlightenment

**Richard Colmorn and Michael Hülsmann**

**Abstract** In the idea contest for technological solutions, cost-efficient processes and customer-oriented mobility services for the usage of electric mobility the phase of a realistic estimation concerning the threats and opportunities can be identified with the help of the hype cycle. Therefore, the research question arises about the steps which lead towards the slope of enlightenment. In this regard, the paper intends to discuss the questions in the following categories for getting an extensive overview: How did the market structures change? How did the business models change? How will the service strategies change? As a result, the paper contributes to the research about future concepts of mobility by providing an overview of different strategic perspectives.

**Keywords** Electric mobility · Market structures · Business models

## 1 The Slope of Enlightenment

In the headings of German online newspapers a change in the expectations about the success of electric mobility can be observed in the course of time. For example, optimistic headlines in 2009 and 2010 such as "Infected by the electric fever" (Pander 2009) and "In the intoxication of electrodes" (Rother 2010) can be distinguished from sceptical ones in 2011 such as "No fast successes in the development of batteries are expected to arise" (Handelsblatt Online 2011) and "Electric cars become accepted more slowly than expected" (Eckl-Dorna 2011) and pessimistic ones, e.g. "The electric car is going to die" (Höltmann 2012) and "It is a long march" (Süddeutsche 2012). This development of expectations or opinions about the success

R. Colmorn (✉) · M. Hülsmann
Jacobs University Bremen gGmbH, Campus Ring 1, 28759 Bremen, Germany
e-mail: r.colmorn@jacobs-university.de

M. Hülsmann
e-mail: m.huelsmann@jacobs-university.de

 
D. Fornahl and M. Hülsmann (eds.), *Markets and Policy Measures in the Evolution of Electric Mobility*, Lecture Notes in Mobility,
DOI 10.1007/978-3-319-24229-3_2

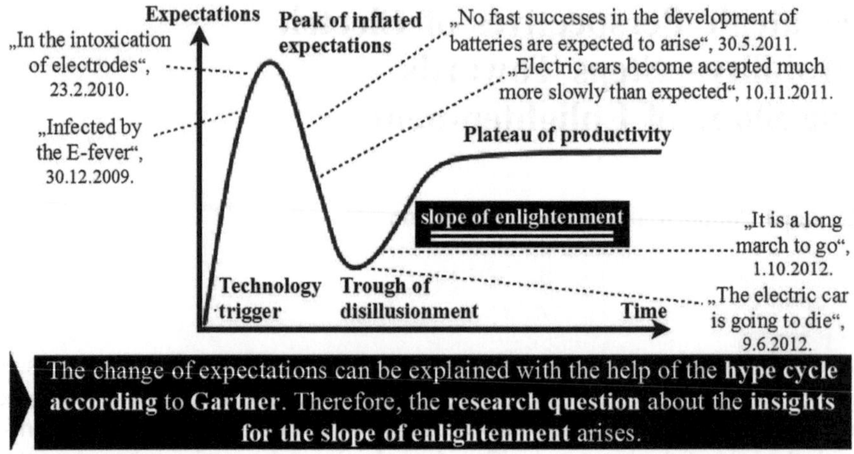


**Fig. 1** Illustration of the change of expectations over time with the help of the hype cycle according to Gartner (cf. N.N. 2014)

of electric mobility can be approached in a first step with the help of the hype cycle according to Gartner (cf. N.N. 2014). The hype cycle according to Gartner is a graphical tool from the advisory firm Gartner Inc. for illustrating the typical phase of the expectations and acceptances of new technologies. According to that, a technology trigger such as the development of electric mobility appears, when no usable concepts are available at the market. A peak of inflated expectations follows because the media jumps on the topic with stories about the potential outcomes. Due to the fact that the efforts in research and development and first applications fail to deliver, negative headlines come more to the front. This is followed by the phase of the slope of enlightenment in which the opportunities and threats as well as the strengths and weaknesses of the new technology have been understood so that a more realistic estimation becomes possible towards the plateau of productivity in which a market penetration due to the fulfilment of the customer needs arises. By assigning the above-mentioned exemplary headlines to the typical curve of a hype cycle, the development of the change of opinions and expectations about electric mobility can be explained. Therefore, the **research question** can be derived about the insights for the slope of enlightenment so that the preconditions for the market penetration can be acknowledged (Fig. 1).

The research question refers to the scientific discipline of Strategic Management because it deals with the question—inter alia—about how to be successful on the market in the long run (Reisinger et al. 2013). For doing so, it is important to identify the opportunities and threats in the business environment that is becoming more and more complex and dynamic. For this, the paper shall help to get the "big picture" when talking about electric mobility so that the following considerations orientate themselves to the typical proceeding of a strategic analysis. For example, Reisinger et al. (2013) distinguish between an internal (comprising the structure, culture and strategies) and an external strategic triangular (comprising the positioning and the distribution and strategies) so that they are connected through the sets of strategies for doing a strategic analysis. By considering the constrains of this paper, Sect. 2 will consider the question about the changes in the market structures for the external triangular because this allows to identify findings in the company's environment, while Sect. 3 will focus on changes of the business models for the internal triangular because of their focus on the configurations of the value creation of a company. Section 4 will consider the question about changes in the service strategies for combining the external and internal perspective before Sect. 5 can conclude with a summary towards the main findings for the slope of enlightenment.

## 2 How Did the Market Structures Change?

Starting point for the investigation of the strategic perspectives was the question about the relevant external influences for the development of electric mobility because insights about these influential factors allow to derive conclusions of so-called enabling or disabling interrelations from the environment. For this purpose, a database that had been started in a previous project was continued. In this context, online-articles were identified, systematically saved and analysed with regard to the keyword of "Elektromobilität" (German for electric mobility). The systematic approach for saving the articles from Spiegel-Online (including Manager Magazin), Wirtschaftswoche, Süddeutsche and the Zeit was the result of the date in the type of "yyyy-mm-tt-" plus the automatically suggested name for the pdf by saving the article. For the analysis of these articles, different categories were used. The first category examined, if a positive, neutral or negative impression could be identified, of course, from a subjective point of view. For indicating these trends, the first category was indicated through "+1", "0" or "−1". Thus, it became possible to graphically and statistically investigate the opinions about the electric mobility over time because the dates could be transformed to a time scale, while the values could be scaled in a mathematically cumulative way. The following four categories corresponded to political-legal, economic, socio-cultural and technological factors of a so-called PEST-analysis. In general, it is aim of a PEST-analysis to identify influential factors in the four above-mentioned categories that can directly or indirectly affect the organisation so the potentials of success are supported or weakened (cf. e.g. Kerth et al. 2009). With the help of this database and the statistical analysis of the cumulative values in the "trend-category", the hypothesis about a hype cycle of the expectation about electric mobility could be further supported. Doing so, the advantages and disadvantages of this technology are more realistically integrated. Additionally, the identified influential factors could also be clustered with regard to their statement of being an opportunity or a threat for the development of Electric Mobility. While a more detailed publication of this extensive analysis is in preparation, at this stage the general conclusion can be drawn that opportunities and threats can be identified in equal measure. Therefore, a final conclusion cannot be drawn so that the environment becomes more complex and dynamic but the dominating trends are hard to determine.

As another important aspect for figuring out the changes in the market structures, the question—inter alia—about the competing forces can be addressed. The reason for this is that changes in the industry-wide competition are generally connected with changes in the market shares and companies successes (Kerth et al. 2009). Porter's Five Forces are a well-established approach for analysing the external competing situation of an industry (Porter and Brandt 2009). It is based on the idea that the market attractiveness is the result of the intensity of the competitive rivalry, i.e. the barriers to market entry. This intensity of competitive rivalry is the result of the bargaining power of suppliers and customers as well as the threats of potential customers and of substitute products.

The **threat of substitute products** expresses the idea that the customer needs or requirements can generally be fulfilled by another product or service, too, so that the product or service is substituted. In the context of electric mobility a threat of substitution exists because different driving technologies exist respectively are being developed such as traditional cars on the basis of fossil fuels, fuel cell vehicles or hybrid vehicles such as Plug-In-Hybrids or Range Extenders (cf. Bundesregierung 2009). Therefore, an increasing competition can be assumed but considering the amount of e.g. hybrid vehicles that are currently available on the German market it can be seen that the German car manufacturers as one major player for the development of electric mobility are well positioned (Fig. 2).

These considerations are quite similar to the ones about the **threat of new competitors** dealing with the market entry of new companies. In this regard it can be observed that most of the car manufacturers extend their product portfolio by developing electric vehicles. Hence, only an increase of the competition is expected to arise in the short run because new companies such as Stromos E Cars, Think or Move About have already gone bankrupt so that only Tesla Motors and BYD promise to have a successful market position in the long run (Table 1).

With regard to the bargaining powers it can be argued that the **bargaining power of customers** will further increase because the customer preferences and their expectations will further increase such as e.g. with respect to the greenness or IT-based functionalities of the car. At the same time, the willingness to pay decreases while only certain customer segments concentrate on the brand image. The **bargaining power of the suppliers** has increased, too, because the trend of a shift of the value creation towards the supplier will further continue due to increasing requirements of research and development of high-technology products. In addition to these competencies, the availability of resources plays an essential part. For example, for the production of e-motors the materials of iron, neodymium, chopper and noble earths are essential that are only produced in certain—often politically instable—regions. Besides a price increase of these raw materials due to an increase of demand a stronger dependence on few suppliers arises. Finally, the removal of existing and the adding of new components will support this trend.

In consequence, it can be summarized that the competing forces within the industry have increased and it is expected that they will continue to increase in the

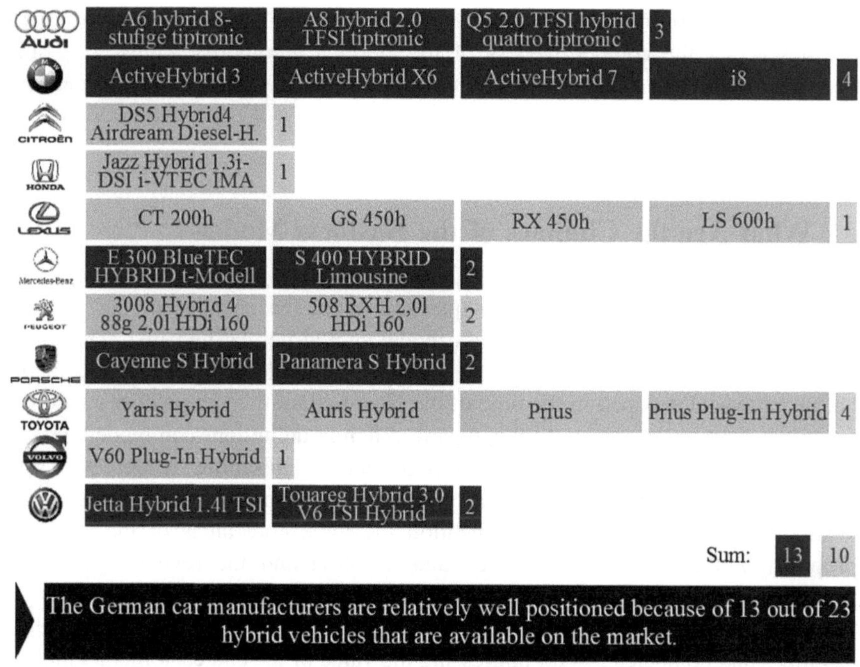


**Fig. 2** Illustration of the threat of potential substitutes through e.g. hybrid vehicles

**Table 1** Battery electric vehicles available on the German market (as per April 2014)

| Brand and model | List price (€) | Range (km) |
|---|---|---|
| BMW i3 | 34.950 | 171 |
| Chevrolet Volt | 42.950 | 95 |
| Citroen C-Zero | 29.393 | 119 |
| Ford Focus Electric | 39.990 | 149 |
| German E-Cars Stromos | 25.950 | 101 |
| Mercedes-Benz SLS | 416.500 | 250 |
| Mia Electric | 25.504 | 64 |
| Mitsubishi iMiEV | 23.790 | 119 |
| Nissan Leaf | 29.690 | 160 |
| Peugeot iOn | 23.393 | 119 |
| Renault ZOE | 21.700 | 151 |
| Smart Fortwo | 23.680 | 132 |
| Tesla Model S | 76.740 | 500 |
| Volkswagen e-Golf | 34.900 | 147 |
| Volkswagen eUp | 26.900 | 160 |
| 15 models including six German ones | Ø 32.824[a] | Ø 156[a] |

[a]Without Mercedes-Benz because of its high deviation

mid- and long-term (cf. Colmorn and Hülsmann 2013). Nevertheless, it could also be shown that the competitiveness of the German automotive industry is quite promising, which also has positive implications for the development of electric mobility.

## 3 What Are the Changes of the Business Models?

In order to answer the overarching question about the changes of the business models, the terminology of a business model should be clarified first. The terminology of a business model can be assigned to the field of business informatics because it was developed in the time of the New Economy to show how a business functions in an abstract way. In this regard, different definitions can be defined e.g. by highlighting customers, offerings, activities and organisations, resources, supply of factors, production inputs or the scope of management, while for example Wallentowitz et al. (2009, 2010) mention the three core areas of the customer proposition, the architecture of the value creation and the revenue generation models. Nevertheless, it can be assumed that independent of the exact definition a business model generally tries to determine how a company can configure its resources and competencies for addressing the value of the end customer so that the questions about their changes seem important because of internal changes in a company.

In the scientific literature and the daily news, new forms of business models are discussed. Therefore, it becomes possible to categorize business models, considering whether they constitute archetypes of business models or new forms. While traditional forms can generally be distinguished from each other with regard to their scope of costs (total costs vs. product costs) and their scales of usage (fixed vs. usage-dependent), new forms of business models cannot clearly be identified (Fig. 3).

To what extent these business models are competitive for electric mobility has been investigated in Colmorn and Hülsmann (2014). In that paper, a Total Cost of Ownership-approach (cf. e.g. Carr and Ittner 1992) was used for determining the customer value with regard to the Total Costs in dependence on the price development and the yearly driving performance. An empirical database was collected containing information about prices as well as usage-dependent and usage-independent costs e.g. for the Volkswagen Golf VII 1.4 TSI and the eGolf so that the two driving technologies could be compared with each other. As a result, it could be shown that independent of the different price developments for electric energy and fossil fuels it is assumed that battery electric vehicles will not be competitive in the middle term due to the high purchase costs for the battery. Therefore, it can be derived that business models for the IT-based services might be a more promising approach in the context of future mobility concepts because they can be used independently of the driving technologies that will be successful on the market in the long run.

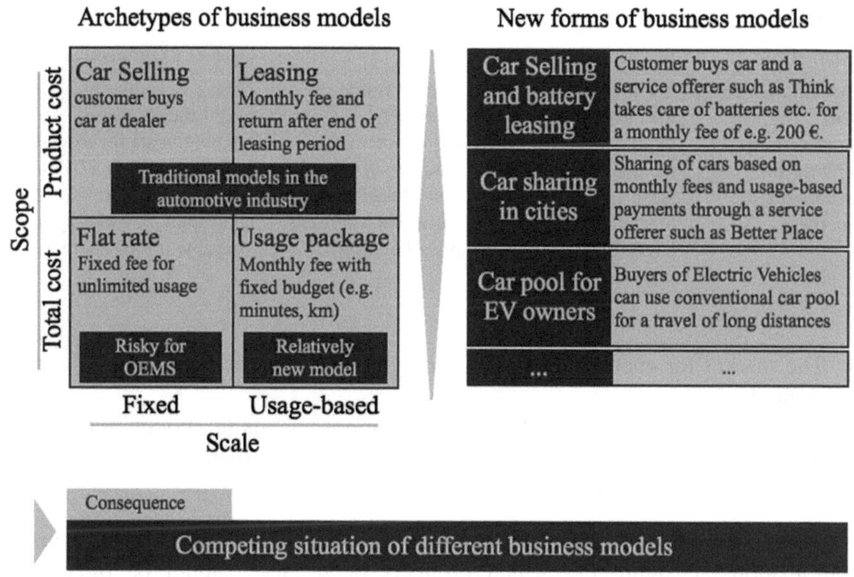


**Fig. 3** Traditional and new forms of business models (from Colmorn 2014 with reference to McKinseay 2010)

## 4 How Will the Service Strategies Change?

Service strategies help to realize the value proposition for the end-customer so that the diffusion of electric mobility is supported and the risk of market failure can be reduced. Thus, service strategies can be understood as a goal-oriented end-means-combination to be successful on the market by integrating the results of the internal and external perspective. In order to do so, this section intends to investigate the market positioning and the respective observable customer benefit as one option for answering the question about the service strategies.

This approach is based on the assumption that the technology of electric mobility does not focus on the ownership of a special product any longer; instead of this the offering of a mobility service becomes more and more important. In consequence, it seems prudent to specify the potential markets for electric mobility because of this change to a more service-oriented character for mobility. Thus, this problem can be directly traced back to the result of the PEST-analysis because the functionalities for the usage as well as the increased customer requirements were identified as a significant change in the market structures. In combination with the results of the economic as well as technological factors, the following markets and their corresponding market segments can be identified:

- The market for the electric cars is the result of the three segments of the chassis, the electric motor/powertrain and the battery/storage system.

**Table 2** Illustration of the market segments

| Market for... | Electric cars | Energy | Car-2-X-communications |
|---|---|---|---|
| Assumption for the content-based criteria for separating the markets | Electric cars fulfil the customer requirements for reliable, secure and clean mean of transport at the payable price-benefit-ratio | Energy fulfils the customer requirements for climate-saving and easy charging of the battery at a payable price-benefit-ratio | Car-2-X-communications fulfil the customer requirements for extended opportunities of IT-based services for increasing the efficiency of mobility at a payable price-benefit-ratio |

- The market for energy comprises the segment of electric energy and charging technologies.
- The third market for “car-2-x-communications” is more service-oriented by focussing on the interfaces with other cars, the infrastructure, mobile devices, internet, home/office and enterprises.

Based on potential growth strategies that focus on the product-market-combinations to figure out which products or services should be offered on which market, assumptions for content-based criteria for separating the markets could be derived with the help of the results of the changes in the markets structures as summarized in Table 2.

Despite of the fact that further criteria for separating the markets from each other can be found in the scientific literature of marketing research, a positioning problem can be generally derived because different options exist about how to strategically behave. This points to the theoretical approaches for realising competitive advantages because by considering the strengths of the company and the situation of the competitors a competitive advantage is defined respectively perceived by the end-customer pointing to the concept of the customer value. Thus, it becomes possible to use the analytical approaches of the generic strategies according to Porter and Brandt (2009). These generic strategies constitute that companies generally have two options for addressing the customer value, i.e. the selection of the strategy and the selection of the place of competition. Based on this, four option for the competitive positioning can be derived.

- Cost leader—core market
- Cost leader—niche market
- Differentiation—core market
- Differentiation—niche market

With the help of secondary literature from the discipline of Strategic Management a setting of success factors could be identified for realizing the strategy of cost leadership (e.g. size-dependent cost degressions, learning effects, degree of capacity utilisation, cross-functional interrelations and integrations, cooperation,...) and for the strategy of differentiation (e.g. special product-characteristics, dealer network, potentials for innovation, marketing activities,...)

so that potentials of success for the market positioning can be derived. Doing so, our research discussed the contributions and limitations of the strategies for the positioning on the markets that can only be represented in a compressed form (Table 3).

With the help of these generic competitive strategies according to Porter and Brandt (2009) different models for positioning on the market can be derived. For example, the capability to adapt to special customer requirements and the capability to generally solve problems opens up possibilities for positioning as exemplary illustrated using the example of the changes in the market structures.

**Table 3** Discussion of the types of strategies for the different markets

| | Car market | Energy market | Market for Car-2-X communications |
|---|---|---|---|
| Niche versus core market (contributions) | Market culmination in the beginning—later niche is possible | Electric service provider: core market | Core markets for systems requiring new developments |
| | | Offerer of charging stations: core and niche markets are possible | Potential niche strategies for completely new functionalities |
| Cost leadership (contributions) | Taking advantage of available capacities and resources for lowering the prices | Realization of low costs for the customers | Better production utilisations for established systems |
| | | | Adaption to sourcing strategies (cost focus) |
| Cost leadership (limitations) | • Price competition<br>• Neglecting necessary efforts for research and development | • Price competition<br>• In case of external procurement of electric energy—no economies of scope | • Price competition<br>• High investment costs at the beginning |
| Differentiation (contributions) | Price competition can be reduced with the help of special service/product characteristics | Offer of additional services becomes possible for electricity companies | For systems that have to be newly developed, differentiations become possible due to new functionalities or value addeds |
| Differentiation (limitations) | • Basic price is already high so that costs for development can be hardly factored into the price<br>• Exact customer preferences are unknown; fail developments are possible | Additional services can be easily copied through the competitors | • High investments costs at the beginning<br>• Exact customer preferences are unknown; fail developments are possible |

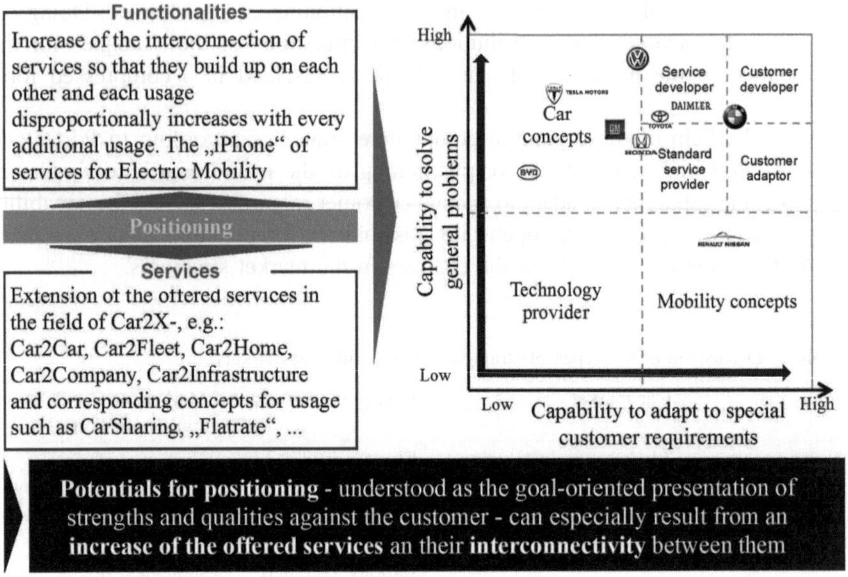


**Fig. 4** Illustration of promising models for positioning on the market

Consequently, with regard to the question about the changes in the service strategies two key insights can be derived: On the one side, new strategic groups for the positioning on the market will arise due to new service-based markets, e.g. in the segment of the Car-2-X-communication. Therefore, from a practical perspective the strategic behaviour will further change and adapt to new characteristics of electric mobility. On the other side, the “rules of the game” stay the same because the framework these strategies had been derived from originate from Porter’s generic competitive strategies. Thus, from a theoretical perspective the analytic frameworks to talk about electric mobility keep their validity (Fig. 4).

## 5 Concluding Remarks

The paper contributed to the research about findings towards the slope of enlightenment for the development of electric mobility by following the three different perspectives of the Strategic Management of the external and internal strategic triangular as well as strategies as the connecting element between them.

With regard to the question how the market structures have changed, it can be concluded that the complexity and dynamic of the environmental influences have increased and will increase further. By considering this complex set of influencing factors they could be assigned to the categories of opportunities and threats for the development of electric mobility. Nevertheless, a reliable statement about which trend is going to occur cannot be drawn at this stage.

With respect to the question about the changes in and of the business models, it was shown that the traditional forms will further exist but that new forms of business models are expected to arise. Hence, it was argued that it is still unclear which of the new forms are the most promising ones or how they should be modified. Due to the fact that the traditional forms are not competitive because of high purchase costs—for the battery—it can be assumed that future potential lies in the development of special forms for IT-based services.

Finally, with respect to the strategy services it could be shown that the German car manufacturers are well positioned and that future potentials exist with regard to the development of new services and their interconnection with each other so that each additional usage of one further service increases the economies of scope for the end customer.

As a result, an increase of the strategic complexity can be concluded because of the higher diversity of influencing factors and potential business models or IT-based service strategies. Hence, it is assumed that this increased complexity will result in limited changes in the form of strategic behaviour because the "rules of the games" stays the same (cf. Colmorn and Hülsmann 2013). Therefore, with regard to the question about insights for the slope of enlightenment it can be summarized in a very generalized way that the "big picture" about the chances of success of electric mobility is still unclear so that it is a long march till the plateau of productivity.

**Acknowledgments** This study was kindly funded by the German Federal Ministry of Transport and Digital Infrastructure (BMVI) whose support made this research possible.

## References

Carr L, Ittner C (1992) Measuring the cost of ownership. J Cost Manag 6(3):42–51

Colmorn R, Hülsmann M (2013) Strategic perspectives for electric mobility: some considerations about the automotive industry. In: Hülsmann M, Fornahl D (eds) Evolutionary paths towards the mobility patterns of the future. Springer, Berlin, pp 155–168

Colmorn R, Hülsmann M (2014) Facing the complexity for future concepts of mobility: the total cost of ownership-approach for the evaluation of business models. In: Fornahl D, Hülsmann M (eds) Electric mobility evolution: theoretical, empirical and political aspects. Springer, Berlin

Die Bundesregierung (2009) Nationaler Entwicklungsplan der Bundesregierung. Berlin. http://www.bmvbs.de/SharedDocs/DE/Artikel/UI/elektromobilitaet-nationaler-entwicklungsplan.html?linkToOverview=js. Accessed 09 Oct 2012

Eckl-Dorna W (2011) Elektroautos setzen sich langsamer als erwartet durch [Homepage of manager magazin], [Online]. http://www.manager-magazin.de/unternehmen/autoindustrie/0,2828,796768,00.html. Accessed 10 Nov 2011

Handelsblatt Online (2011) Keine schnellen Batterie-Erfolge in Sicht [Homepage of Wirtschaftswoche], [Online]. http://www.wiwo.de/unternehmen/elektro-autos-keine-schnellen-batterie-erfolge-in-sicht.html. Accessed 21 Nov 2011

Höltmann I (2012) Das E-Auto ist dabei, zu sterben [Homepage of Wirtschaftswoche], [Online]. http://www.wiwo.de/technologie/auto/autoderzukunft/elektroautos...html. Accessed 09 June 2012

Kerth K, Asum H, Stich V (2009) Die besten Strategietools in der Praxis: Welche Werkzeuge brauche ich wann? Wie wende ich sie an? Wo liegen die Grenzen?, 4th edn. Carl Hanser Verlag, München

McKinsey (2010) Vortrag auf der NOW-Konferenz "Clean Mobility Insights - Batterie und Brennstoffzelle als Zukunftsmotor", 15–16 Sept 2010, Berlin, Germany

NN (2014) Hype-Zyklus [Homepage of wikipedia], [Online]. http://de.wikipedia.org/wiki/Hype-Zyklus. Accessed 30 June 2014

Pander J (2009) Angesteckt vom E-Fieber [Homepage of Spiegel Online], [Online]. http://www.spiegel.de/auto/aktuell/revolution-der-autotechnik-angesteckt-vom-e-fieber-a-669364-druck.html. Accessed 30 Dec 2009

Porter M, Brandt V (2009) Wettbewerbsstrategie: Methoden zur Analyse von Branchen und Konkurrenten. Campus Verlag, Frankfurt a.M

Reisinger S, Gattringer R, Strehl F (2013) Strategisches Management, 1st edn. Pearson, Hallbergmoos

Rother F (2010) Elektroauto-Batterien: Im Rausch der Elektroden [Homepage of Wirtschaftswoche], [Online]. http://www.wiwo.de/unternehmen-maerkte/elektroauto-batterien.html. Accessed 23 Feb 2010

Süddeutsche (2012) Es ist ein langer Marsch [Homepage of süddeutsche Zeitung], [Online]. http://www.sueddeutsche.de/auto/elektroauto-gipfel-im-kanzleramt-es-ist-ein-langer-marsch-1.1483527. Accessed 01 Oct 2012

Wallentowitz H, Freialdenhoven A, Olschewski I (2009) Strategien in der Automobilindustrie - Technologietrends und Marktentwicklungen, 1st edn. Vieweg + Teubner, Wiesbaden

Wallentowitz H, Freialdenhoven A, Olschewski I (2010) Strategien zur Elektrifizierung des Antriebsstranges – Technologien, Märkte und Implikationen, 1st edn. Vieweg + Teubner, Wiesbaden

## Author Biographies

**Richard Colmorn** studied physics with special focus on Complex Systems and Econophysics. He was a research associate in the working group "Systems Management" at Jacobs University Bremen. His research interests comprise the Strategic Management of the technological change in the value-adding networks of the automotive industry from a complexity-based perspective.

**Michael Hülsmann** is Managing Director at Jacobs University Bremen gGmbH since June 2014. From 2009 until 2014 he led the department "System Management" at Jacobs University. His main research focus lies on a Strategic Management of Logistics Systems. Thereby, Prof. Hülsmann is doing research mainly in the fields of competence and technology based positioning, accomplishing of technological change in value-adding networks and setting-up of inter-organizational coordination mechanisms.

# On the Road to an Electric Mobility Mass Market—How Can Early Adopters be Characterized?

**Axel Ensslen, Alexandra-Gwyn Paetz, Sonja Babrowski, Patrick Jochem and Wolf Fichtner**

**Abstract** Different field trials and corresponding acceptance studies with new technologies were carried out between 2010 and 2013 at the Chair of Energy Economics at the Karlsruhe Institute of Technology (KIT). Those involved Electric Vehicle (EV) users, Liquefied Petroleum Gas (LPG) and Compressed Natural Gas (CNG) vehicle users as well as persons with strong interest in EV and smart energy home technologies. In order to characterize early adopters the same item-sets concerning attitudes regarding climate change, prices and innovations as well as corresponding socio-demographic characteristics were used throughout all these studies. Survey datasets originating from these studies are joined to be analyzed together. Regression methods are applied in order to characterize early EV adopters based on a subsample of EV company car users in the French-German context. A binary logistic regression model explaining private EV purchase intentions is developed. According to this model, early private EV adopters are likely to have a higher level of income, to have a household equipped with two or more cars and to travel more than 50 km a day, not necessarily by car. This model additionally shows that possibilities to experience EV (e.g. by test drives) are important leverages to support adoption of EV by private car buyers. Respondents who already decided to privately purchase an EV show significantly lower general price sensitivities than the LPG and CNG vehicle users.

---

A. Ensslen (✉) · A.-G. Paetz · S. Babrowski · P. Jochem · W. Fichtner
Institute for Industrial Production – Chair of Energy Economics,
Karlsruhe Institute of Technology (KIT), Hertzstr. 16, 76187 Karlsruhe, Germany
e-mail: axel.ensslen@kit.edu

A.-G. Paetz
e-mail: alexandra-gwyn.paetz@kit.edu

S. Babrowski
e-mail: sonja.babrowski@kit.edu

P. Jochem
e-mail: jochem@kit.edu

W. Fichtner
e-mail: wolf.fichtner@kit.edu

 
D. Fornahl and M. Hülsmann (eds.), *Markets and Policy Measures in the Evolution of Electric Mobility*, Lecture Notes in Mobility,
DOI 10.1007/978-3-319-24229-3_3

**Keywords** Electric vehicle · Electric mobility · Early adopter · User acceptance · User experience · EV diffusion

## 1 Introduction

The European aim of reducing Greenhouse Gas Emissions (GHG) by 80 % by the year 2050 compared to 1990 (European Commission 2011) will cause changes in the transportation sector as today it accounts for about 19 % of the total European GHG emissions (Eurostat 2013a), with a continuously increasing share. As individual road transportation is responsible for the main share of those emissions (Eurostat 2013a), significant changes seem unavoidable with regard to the share of passenger cars running on alternative fuels (cf. Kay et al. 2013).

Vehicles running on liquefied petroleum gas (LPG), compressed natural gas (CNG), and on electricity (EV[1]) have been discussed as a more energy-efficient and climate-friendly means of individual transportation. Cars running on LPG and CNG have been on the market for several years and 500,867 (LPG) respectively 79,065 (CNG) were on German roads by January 2014 (KBA 2014). Even though those two technologies have the highest share among cars with alternative fuels, LPG and CNG cars account for only about 1 % of the total German passenger car fleet (KBA 2014).

Due to positive developments in the battery technology (Thielmann et al. 2012), battery electric vehicles (BEV) have undertaken a rebirth in the last years. Currently around 40 different EV are offered on the German market (Eckl-Dorna and Sorge 2013) and 12,156 cars were registered on January 1st, 2014 (KBA 2014). Although the market seems dynamic (on a low level), the market stage is somewhat earlier compared to LPG and CNG. At the same time German policy measures are strongly aiming at entering the mass market with EV. Accordingly, the government pronounced the ambitious targets of 1 Million EV in 2020 and 6 Million EV in 2030 (BMVBS 2011). Assuming that the German passenger car fleet remains constant, that means that about 2.5 % respectively 15 % of the passenger cars would be substituted with EV. Even though policy measures have predominantly been targeting to technology developments that are supportive to the supply side of four-wheeled EV (cf. Bundesregierung 2009), a strong market penetration of electric two-wheelers can already be observed. Over one million pedal electric cycles (pedelecs) and electric scooters (e-scooters) are already on German roads (Dütschke et al. 2013).

[1]EV is used as synonym for all vehicles including Battery Electric Vehicles or All-Electric Vehicles (BEV), Range Extended Electric Vehicles (REEV) and Plug-in Hybrid Electric Vehicles (PHEV).

Policy measures are needed in order to extend the success of electric drives from pedelecs and e-scooters to passenger cars capable of being charged in a smart way. But at whom should the measures supportive to EV diffusion target? Who are the first potential buyers of EV and how can they be characterized? Do these early EV adopters resemble early adopters of other technologies, such as LPG/CNG and smart energy home technologies? If so, can we learn from those market experiences?

Smart energy home technologies are particularly focused on due to the fact that higher penetration rates of EV will have an increasing impact on the electricity system (cf. Jochem et al. 2013). A higher share of EV in the car fleet will increase power demand during peak hours in residential areas considerably, if EV are not charged in a controlled "smart" way. Possible solutions to this challenge could be provided by demand side management solutions, such as automatic delayed EV charging (cf. Jochem et al. 2013). The technical solutions needed to make smart charging possible are based on Information and Communication Technologies (ICT) and discussed as smart technologies, such as charging spots with smart meters.

In this article early adopters are characterized as individuals who have already been using or are actively interested in innovations and are therefore likely to adopt these innovations when the markets reach early adopter phase. Up to now, this has not been the case for the markets of smart energy home technologies (cf. Bundesnetzagentur 2011) and EV (cf. Wietschel et al. 2013). The EV market is in a very early phase, i.e. only innovators already took the decision to purchase an EV in Germany. Unlike that, LPG and CNG early adopters already took the purchase decision, as this market is in an advanced position compared to the market of EV and smart energy home technologies (cf. Dütschke et al. 2011).

In order to characterize innovators and early adopters of EV the authors compare innovators and early adopters from different studies considering their price sensitivities, their environmental awareness as well as their innovativeness. Furthermore, the authors look at early EV adopters' attitudes towards EV and try to identify their willingness to purchase an EV according to their mobility behavior, their experience levels with EV as well as socio-demographic characteristics.

This article has the following structure. A literature review (2) is conducted before the survey subsamples under consideration are presented and characterized by analyzing corresponding environmental awareness, price sensitivities, innovativeness as well as socio-demographic characteristics (3). Furthermore, regression approaches are applied to explain potential willingness to privately purchase an EV within the next years based on fleet EV users' attitudes and norms, their mobility behaviors, their experience levels with EV as well as their socio-demographic backgrounds (4). After a brief summary and conclusion the outlook discusses how one of the models could be applied to the existing representative mobility studies (e.g. ENTD,[2] MiD[3]) in order to derive conclusions about EV adoption potentials within the next years in France and Germany (5).

# 2 Literature Review

## 2.1 Theory on Diffusion of Innovations

According to Rogers (2003), the diffusion of new products such as EV takes place in several steps depending on the share of customers that adopt the new product over time. In the first stage, so called innovators try these new products as soon as they are offered on the market. They usually make up for about 2.5 % of the population. Assuming that all vehicles on German roads (43.9 Million, cf. KBA 2014) would be replaced with EV, the main users of somewhat more than one million EV would all be called innovators. Afterwards early adopters follow (13.5 %). Then the early and late majority (34 % each) and at last the so called laggards (16 %) who are not really interested in new products follow.

As adoption can be described as decision making process of individuals to finally accept an innovation, using stated preference survey data in order to identify the early adopters of an innovation seems appropriate (cf. Bass 2004)

## 2.2 Characterizing Early EV Adopters Based on Stated Preference Survey Data

Wietschel et al. (2012) identify early adopters of EV in Germany until 2020 on the basis of surveys and group discussions with EV users focusing on their economic, attitudinal and socio-demographic backgrounds. They indicate that the probability of privately purchasing an EV among current users is highest for men in the beginning of their 40s, with a higher socio-economic status and most likely having a technical profession. This potential customer group is likely to live in multi-person households with several vehicles, which tend to be in rural areas or in the outskirts. However, selling EV only to this group will not be sufficient in order to target one million EV until 2020. According to Wietschel et al. (2012) about 50,000 vehicles could be sold to this group annually. About 80,000 vehicles would be needed to be sold annually to private customers to reach the German goal of one

[2]Enquête nationale transports et déplacements 2008.

[3]Mobilität in Deutschland 2008.

million EV. In order to derive conclusions about the diffusion process Wietschel et al. (2013) characterize innovators as EV users (0.5 % of the population), early adopters as individuals interested in EV with purchase intention (1 % of the population) and the early majority as individuals without purchase intention but interested in EV (48 % of the population). The remaining share of the population are identified as laggards and as such not interested in EV today.

Hackbarth and Madlener (2013) conduct a discrete choice analysis based on survey data they collected in Germany. They applied a mixed logit model in order to derive conclusions about potential demand for Alternative Fuel Vehicles (AFV) in Germany, particularly for plug-in cars. Results indicate that relatively young, well-educated and environmentally-aware survey participants who have the possibility to plug in their car at home and undertake numerous urban trips are most sensitive to AFV adoption. Opposing the findings of Mabit and Fosgerau (2011) and Ziegler (2012) the model of Hackbarth and Madlener (2013) explaining EV purchase intentions does not observe significant influences of the variables gender, number of children and number of cars in the household.

Glerum et al. (2013) forecast EV demand by accounting for attitudes and perceptions. Their analyses are based on stated preference surveys with personalized choice situations involving conventional cars and EV. Swiss survey participants were interviewed at the beginning of 2011. They characterize target EV customers as public transportation users living in households owning several cars, with high incomes and rather young. Furthermore, they find that the introduction of a large incentive (5,000 CHF) on the purchase price of an EV can promote its choice, whereas too-high operating costs (5.40 CHF/100 km) can discourage it.

Ensslen et al. (2012) point out that a quite high number of EV users participating in the fleet test CROME could envision purchasing an EV within the next ten years. Less than 20 % stated not to be willing to do so, about 35 % stated being willing to do so and about 45 % of the respondents were undecided. According to Ensslen et al. (2013a) potential early EV adopters are likely to live in rather rural French areas due to favorable total cost of ownership (TCO), a relaxed parking situation in small municipalities and a high average number of cars per household, which compensates for the range-specific disadvantages most EV have. Annual car mileage is on average higher for people living in small municipalities, which makes TCO favorable. Additionally, French adopters benefit from EV purchase incentives. By the time the survey took place a bonus of 7,000 € was provided by the French government. Furthermore, French adopters benefit from comparably lower electricity costs (cf. Eurostat 2013b) which additionally improve TCO calculations for French EV adopters. After the EV users were experiencing the EV for about a year, user acceptance was studied with a second survey. Ensslen et al. (2013b) analyze BEV users' attitudes and norms potentially influencing BEV purchase decisions in the French-German context. The authors stress that French BEV users are more concerned about climate change than their German counterparts. Furthermore, their results show, that the French respondents indicate a higher innovativeness level. On the other hand the German respondents indicate to a higher degree that having BEV as company cars has a positive external communication effect. Also, highly significant differences can be observed concerning the French and German BEV users' degrees of satisfaction with the BEVs' $CO_2$ emission characteristics. Although Wietschel et al. (2012) as well as Ensslen et al. (2013a) point out BEVs' advantages in less urbanized areas (due to an easy access to charging possibilities at home and better TCO values), users living in rather urban municipalities indicate higher degrees of satisfaction with different characteristics of BEV (e.g. low $CO_2$ emissions, sufficient range) (cf. Ensslen et al. 2013b).

Peters et al. (2011a) describe that energy-relevant purchase decisions of consumers for a passenger car can to a large extent be explained by psychological factors like attitudes towards more fuel-saving vehicles and awareness of problems related to fuel consumption. Therefore, Peters et al. (2011a) extended Rogers' theoretical framework specifically for EVs and determined an additional dimension called social norm influencing the EV adoption process. Peters et al. (2011b, 2013) base their theoretical framework explaining households' purchase decisions of fuel efficient vehicles in Switzerland on a theoretical model of Bamberg and Möser (2007) which includes psychological factors including social and personal norms, problem awareness and perceived behavioural control.

Dütschke et al. (2011) derive conclusions about policy measures supportive to EV diffusion on survey data received from LPG and CNG adopters collected in the end of the year 2010 from individuals who indeed purchased a LPG or CNG car. According to their results, motives to purchase LPG and CNG cars are rather not likely being linked to an innovative technology. Economic aspects are most important, followed by ecological aspects. Accordingly EVs have a decisive advantage as they have the image to be environmentally friendly. At the point of time of the survey a big disadvantage of EV were their comparably high purchase prices. According to the survey participants information concerning LPG and CNG cars were preferably collected by talking to other users during the decision making process. Dütschke et al. (2011) conclude that strategies like field trials could be supportive for EV diffusion, as they could be contributive to a better perception of the reliability and safety of the EV technology. Several other studies are supportive to the field trial strategy, too, as they mention positive reactions of individuals who have tested and used EV (cf. Peters and Dütschke 2010, Peters and Hoffmann 2011). Peters et al. (2011a, b) recommend providing low-threshold opportunities to test EV to consumers, especially for marketing campaigns and promotional measures, as EVs' driving characteristics are often perceived as very positive.

Frenzel et al. (2015) characterize early German EV adopters based on a large survey sample of 3.111 private and professional EV users. Their results show that private EV users are predominantly highly educated men with higher incomes. On average they are around 51 years old and so older than persons buying new conventional cars. The majority of EV users rather live in small towns and rural areas. Despite a high environmental awareness 80% of the households own a second car, particularly if the EV is an all-electric BEV. Concerning professional EV users Frenzel et al. (2015) show that majorly small organizations with up to 49 employees and up to nine vehicles in the fleet use EV.

Plötz et al. (2014) answer the question who will buy EV in Germany by focusing on two subquestions. These are answered by two different methodological approaches and datasets. On the one hand the question who is willing to buy an EV is analyzed based on survey data from responents with different levels of interest in EV including EV users. On the other hand the question who should buy an EV is analyzed by calculating TCO (c.f. chapter 2.3.) of EV based on individual car mileages provided in a representative German mobility study. Results indicate that private EV buyers in Germany comprise middle-aged men with technical professions living in rural or suburban multi-person households owning a large share of vehicles in general. They tend to profit from the economical benefits of EV due to their annual vehicle kilometers travelled and the share of inner-city driving. They show a comparably high willingness to buy EV and their comparably high socio-economic status allows them to purchase EV.

## 2.3 *Identification of EV Adopters in Representative Datasets*

The early adopters in representative mobility studies are mostly identified by rational choice decisions such as TCO. Emotional and maybe "irrational" reasons for buying an EV are neglected. This subchapter provides a brief overview on studies identifying early EV adopters in representative studies based on rational choice.

Mendes Lopes et al. (2014) identify the households to whom limited range BEV would be a plausible choice based on a rule-based screening methodology. Households in the Lisbon Metropolitan Area in Portugal are classified according to a set of indicators (e.g. home location, daily trips). Five profiles are defined which correspond to an increasing probability of including BEV in their choice set. According to their results, BEV are only suitable for 1.8 % of the households in the Lisbon Metropolitan Area. Existence of financial incentives would increase the share of EV qualifying households to 6.2 %. However, even if prices would not be a barrier, BEV would only be suitable to 10.4 % of the households under consideration. According to their analyses, households with more cars, a higher household income, with adequate parking space and a travelling distance that fits to the limited range of BEV qualify for BEV.

Windisch (2013) conducted an analysis on the effectiveness of monetary demand-side policy measures (e.g. purchase primes) on EV diffusion of privately held vehicles. The author chose a disaggregate approach based on the database of the French National Transport Survey 2007/2008. Results of TCO calculations show that EV, BEV in particular, can be financially interesting to private customers under certain conditions. Furthermore, her results show that long-electric-range PHEV do not appear to be a financially viable alternative under any realistic vehicle usage assumptions. Her results show that around 35 % of the French households are adapted to the needs and limitations of a BEV, i.e. motorized households with access to parking infrastructure where recharge infrastructure could be installed and with vehicle usage behavior not being constrained by BEVs' limited range. Furthermore, her results show that 51 % of the French households are compatible with a PHEV, i.e. motorized households with access to parking infrastructure that can be equipped with a battery recharging infrastructure. Her results even indicate that the need for private parking infrastructure is a more limiting factor to potential EV uptake than compatible EV usage behavior. Concerning the TCO calculations, her results show high sensitivities concerning EV qualifying households according to the purchase primes granted by the government. Under a 5,000 € purchase bonus buying a BEV is only rational for 3.5 % of the French households. A 7,000 € purchase bonus on the other hand makes it rational for about 28.2 % of the households to purchase a BEV. Her results show that in rather rural areas (Petite and Grande Couronne) a lot of the household are practically compatible with BEVs' needs according to their driving patterns (17.3 % and 31.4 %). However, considering the TCO only 1.4 % respectively 3.0 % of the households in the Petite respectively the Grande Couronne area are EV qualified. In the urban center (Paris) on the other hand only 6.9 % of the households are practically compatible with BEVs. However, all of these household would also qualify for BEV from a TCO point of view when preferential parking tariffs for EVs would be introduced in urban centers. Furthermore, the ongoing decreasing battery price would lead to lower required purchase bonus payments as of today.

Wietschel et al. (2013) also base their analysis concerning EV diffusion in Germany on TCO calculations. For private and official car users their analysis was based on data from the German Mobility Panel (MOP 2012), additionally they also consider operational driving profiles of company fleet vehicles (cf. Fraunhofer ISI 2012). Overall 6,500 driving profiles formed the basis for their calculations. Framework conditions are described for three scenarios (pro EV, middle, contra EV) including the development of economic parameters (i.e. fuel prices, battery prices and electricity prices). TCO calculations can be performed considering infrastructure costs, limited supply of adequate EV models as well as increased willingness to pay for an EV. Their results show that depending on the scenario settings EV stock in Germany in 2020 varies between 50,000 and 1,400,000. Their results also indicate that the EV diffusion is sensitive to monetary demand side policy measures (purchase bonus payments up to 2,000 €). EV stock could be almost doubled until 2020 with a purchase prime of only 1,000 €. Main profiteer would be commercial fleets who would make up for almost 60 % of market growth.

Based on a TCO model Pfahl et al. (2013) show that only half of the targeted 1 million EV can be expected by 2020 in Germany without subsidies. They also find that small changes of parameters (e.g. increase of oil price, decrease of battery costs, etc.) can lead to significant higher numbers of EV on the German car market.

# 3 Characterization of Early Adopters

## 3.1 Data Used

In order to gain more information about possible early EV adopters and therefore about the possible diffusion process of EV, several studies that were conducted at KIT's Chair of Energy Economics in the years 2010–2013 focusing on the attitudes of the study participants are joined and analyzed together. This is possible, as a set of several items concerning beliefs and attitudes was used throughout all studies using Roger's theory of diffusion of innovations as conceptual framework (Rogers 2003). The participants differ with regard to their prior experiences and their stage of adoption with the technologies under study: no prior experience (interested, but no adoption), regular users (adoption, but no buying decision), adopters (real buying decision).

The following different technologies are considered: LPG and CNG vehicles, EV, as well as smart energy home technologies. Or more precisely, the following six subgroups are considered (cf. Table 1): (1) private LPG and CNG vehicle users with prior experience and with already accomplished buying decision, (2) survey participants interested in smart energy home technologies, (3) e-scooter users with prior experiences, (4) private EV users with (only) prior experience and already accomplished buying decision, (5) EV company car users with prior experiences, as well as (6) survey participants interested in EV predominantly not having experienced EV. The different groups are briefly characterized among others by their socio-demographic backgrounds.

(1) The respondents grouped in the cluster CNG and LPG vehicle users come from Germany and are about as old as the EV company car users (5). It is noticeable that 95 % of this group's respondents are male. Their level of education is somewhat lower than the educational level of the EV company car users. Furthermore, the LPG/CNG car users' level of income is somewhat lower than the level of income of the EV company car users. These car drivers not only use their LPG/CNG car on a regular basis, but they have also bought these cars some time ago — they are thus customers that accomplished a real buying decision process. This dataset has also been used by Dütschke et al. (2011).

(2) The persons of our sample interested in smart energy home technologies are also predominantly male (70 %). Their level of education is comparable to the EV company car users' as the majority has completed their studies. On the other hand less of them have completed vocational education, but more of them have a high school degree at university entrance level. They are comparatively young and are predominantly living in the region of Karlsruhe. Supposedly many master students have participated in this survey, what would also explain the comparably low levels of income despite their final degrees (i.e. the Bachelor degree).

**Table 1** Overview of the different subgroups within the sample considered in this article

| | | Subsamples considered | | | | | |
|---|---|---|---|---|---|---|---|
| | | (1) | (2) | (3) | (4) | (5) | (6) |
| | | LPG and CNG vehcle users | Persons interested in smart energy home technologies | E-scooter users | Private EV customers | EV company car users | Persons interested in EV |
| Total of respondents (n = 756) | | 130 | 284 | 21 | 17 | 171 | 133 |
| Survey period | | Dec. 2010 | June 2012 | July 2012 and July 2013 | Jan.–Sep. 2011 | Sep. 2012–May 2013 | Dec. 2012–Feb. 2013 |
| Prior experience | | x | | x | x | x | |
| Purchase decision | | x | | | x | | |
| Age | n | 119 | 0 | 11 | 16 | 133 | 133 |
| Arithmetic average | | 43 | n.a. | 23 | 38 | 44 | 37 |
| Gender | n | 120 | 284 | 0 | 13 | 131 | 128 |
| Female | | 5 % | 29 % | n.a. | 15 % | 28 % | 30 % |
| Male | | 95 % | 71 % | n.a. | 85 % | 72 % | 70 % |
| Level of education | n | 120 | 284 | 21 | 15 | 153 | 132 |
| Ph.D./Habilitation (%) | | 3 | 3 | 0 | **33** | 1 | 1 |
| Final degree (%) | | 43 | 55 | 0 | **47** | 63 | 51 |
| Vocational education (%) | | 40 | 10 | 0 | 13 | 14 | 16 |
| High School degree at university entrance level (%) | | 0 | 28 | 100 | 7 | 13 | 17 |
| (General) CSE (%) | | 11 | 2 | 0 | 0 | 8 | 12 |
| No formal certificate (%) | | 1 | 1 | 0 | 0 | 0 | 0 |
| Other (%) | | 2 | 1 | 0 | 0 | 1 | 3 |

(continued)

**Table 1** (continued)

| | | Subsamples considered | | | | | |
|---|---|---|---|---|---|---|---|
| | | (1) | (2) | (3) | (4) | (5) | (6) |
| | | LPG and CNG vehcle users | Persons interested in smart energy home technologies | E-scooter users | Private EV customers | EV company car users | Persons interested in EV |
| Total of respondents (n = 756) | | 130 | 284 | 21 | 17 | 171 | 133 |
| Monthly net household income | n | 104 | 218 | 0 | 14 | 79 | 95 |
| <1,000 € | | 0 % | 11 % | n.a. | 0 % | 0 % | 6 % |
| 1,000–1,999 € | | 11 % | 26 % | n.a. | 22 % | 10 % | 10 % |
| 2,000–2,999 € | | 33 % | 20 % | n.a. | 14 % | 22 % | 27 % |
| 3,000–3,999 € | | 31 % | 16 % | n.a. | 0 % | 32 % | 18 % |
| 4,000–4,999 € | | 11 % | 14 % | n.a. | 14 % | 21 % | 10 % |
| 5,000–5,999 € | | 11 % | 6 % | n.a. | 7 % | 9 % | 12 % |
| 6,000–7,000 € | | 2 % | 5 % | n.a. | 7 % | 2 % | 5 % |
| >7,000 € | | 1 % | 2 % | n.a. | **36 %** | 4 % | 12 % |

(3) The e-scooter users are a group of KIT students with an average age of 23 years. They have used an e-scooter during a field operational test over the course of 5 weeks and were selected by application, thus a strong self-selection effect motivating first movers to apply is assumed (cf. Paetz et al. 2012a, 2013).
(4) The private EV customers originate from the Stuttgart area and can be characterized as real innovators (cf. Rogers 2003), as they have really adopted the new technology (regular use as well as buying decision). It is noticeable, that 36 % of them have a monthly net household income above 7,000 EUR and their educational level is the highest, as 80 % of them have completed their studies (33 % of them at Ph.D. level). They are comparably young as they are on average 38 years old (cf. Paetz and Dütschke 2012).
(5) The major part of the respondents in the group of EV company car users are about 44 years old, live in the Upper-Rhine region in Germany or in Alsace-Lorraine in France and have a high level of education. The majority has studied and a final degree. Their household incomes are comparably high. All of them experienced EV for a longer period of time. At the point of time when responses to the online survey were collected, they had experienced EV on average for about one year within the framework of the CROME project (cf. Ensslen et al. 2013b). This sample has further been increased by a dataset that was used by Paetz et al. (2012b) including 15 German EV company car users (cf. Paetz and Dütschke 2012)
(6) The respondents who are interested in EV but did predominantly not have any experiences with EV at the point of time they were participating in the survey (77 %), come from Germany, are on average 37 years old, are predominantly male and their level of education is at about the same level as the sample's of the persons interested in smart energy home technologies, i.e. most of them have completed their studies. On the other hand their level of income is comparably higher.

## 3.2 Methods Used

In order to derive conclusions about the different groups' (cf. Table 1) environmental awareness, their price sensitivities as well as their innovativeness a principal component analysis (PCA) is applied. This is a statistical procedure used to discover structures and to convert a set of observations of possibly correlated variables into a set of values of linearly uncorrelated variables. According to each respondent's evaluation of a set of statements (items) measured on a symmetric agree-disagree Likert scale different factors and corresponding individual factor scores are derived (cf. Table 3 in the Appendix and Backhaus et al. 2008). In order to compare resulting factor scores of the different groups' attitude levels, Kruskal–Wallis one-way analysis of variance (Kruskal-Wallis ANOVA) is used (cf. IBM SPSS Statistics 2013a; Hartung et al. 2005), as Gaussian distributions cannot be assumed (detailed results of differences between the different groups' respondents' attitudes cf. Table 4 in the Appendix). Kruskal–Wallis ANOVA permits to determine whether the different groups' mean values for the three factors differ significantly. This is the case for all three factors to a highly significant degree ($p < 0.001$ for all three factors.). In order to find out which of the subsamples differ from each other, pairwise comparisons are considered (cf. Dunn 1964; IBM SPSS Statistics 2012).

## 3.3 *Results*

By applying PCA to eight items three factors are derived. One measures the respondents' environmental awareness, one their innovativeness and one their price sensitivities (cf. Table 3 in the Appendix and Figs. 1, 2 and 3). The quality of this factor analysis is mediocre as the Kaiser-Meyer-Olkin criterion of Sampling Adequacy is 0.631 (cf. Backhaus et al. 2008). In order to determine whether the scales that were used to measure the three dimensions mentioned before are internally consistent, Cronbach's Alphas are calculated indicating that the scales measuring the respondents' environmental awareness as well as their innovativeness are indeed interrelated to a sufficiently high degree, whereas internal consistency of the factor price sensitivity is not acceptable. More detailed information about the PCA including Cronbach's Alphas as well as each items' measure of sampling adequacy can be found in Table 3 in the Appendix. Furthermore, adjusted

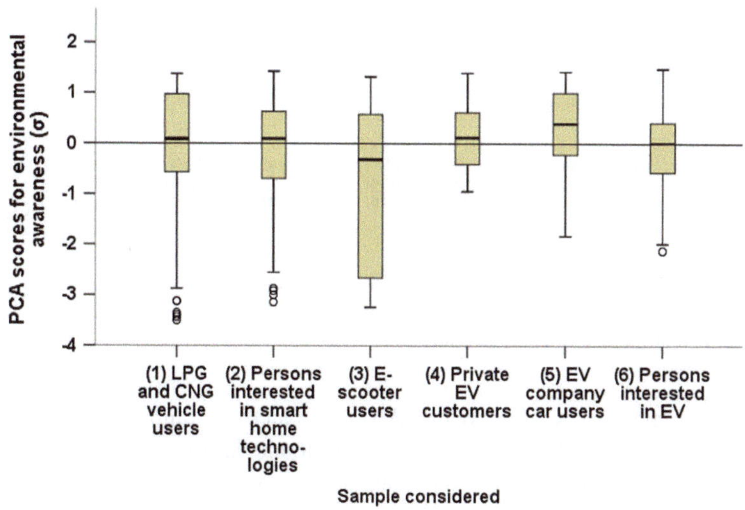


**Fig. 1** Different samples' respondents' environmental awareness

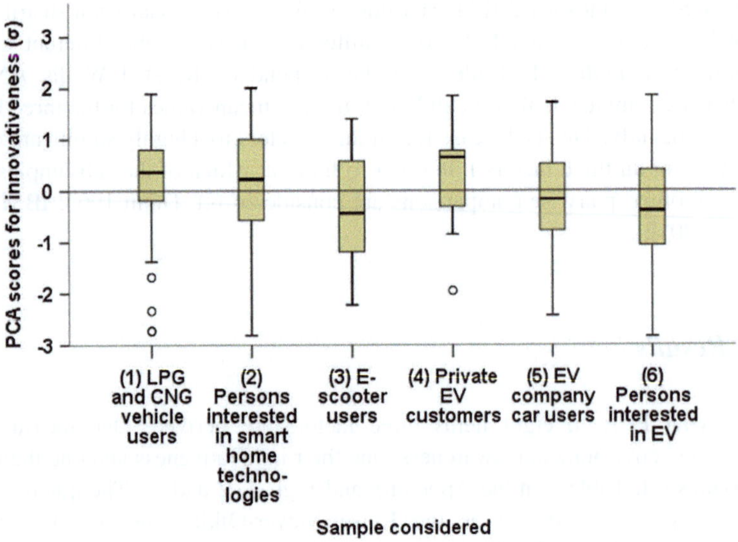


**Fig. 2** Different samples' innovativeness levels

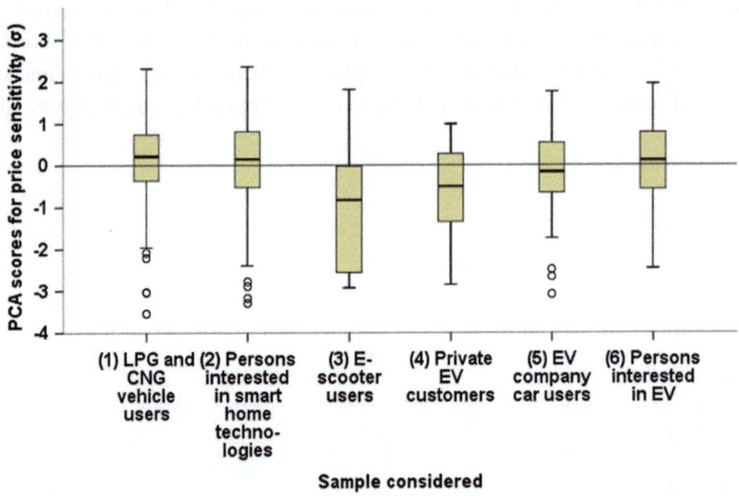


**Fig. 3** Different groups' price sensitivities

significance levels of Kruskal-Wallis ANOVA pairwise comparisons between the factor scores of the different samples can be found in Table 4 in the Appendix. Information on the following boxplot diagrams are available in IBM SPSS Statistics 2013b.

### 3.3.1 Environmental Awareness

Concerning different groups' environmental awareness, significant differences can be observed between French and German EV company car users (5) and almost all other user groups (cf. Fig. 1 and Table 4 in the Appendix). Only private EV customers' (4) environmental awareness is not significantly lower than the EV company car users' (5). As about half of the respondents who are considered in group (5) are French, these differences might mainly be explained by the strong influence the French respondents' had. According to Ensslen et al. (2013b) the French EV users within the CROME project are more worried about climate change than their German counterparts.

### 3.3.2 Innovativeness

Innovativeness of private EV customers (4) and CNG/LPG vehicle users (1) as well as of the respondents who are interested in smart energy home technologies (2) is to a significant degree higher than the innovativeness of those who are interested in EV (6) (cf. Fig. 2). Furthermore, innovativeness of respondents who are interested in smart energy home technologies (2) is significantly higher than innovativeness of respondents who are EV company car users (5).

### 3.3.3 Price Sensitivity

The sample of e-scooter users (3) is to a significant degree less price sensitive than the sample including persons who are interested in EV (6) and to a highly significant degree less price sensitive compared to the respondents who are interested in smart energy home technologies (2) and who are LPG and CNG vehicle users (1). As group (3) mainly consists of students this is highly surprising. However, due to the fact, that the usage of e-scooters is in another price range than the other technologies, the comparison might be biased. According to these findings private EV buyers (4) and e-scooter users (3) are least price sensitive. LPG and CNG vehicle users (1) are significantly more price sensitive as well as the respondents who are interested in smart energy home technologies. Respondents who are interested in EV (6) are to a degree of marginal significance more price sensitive than private EV buyers (4).

## *3.4 Limitations*

Findings concerning price sensitivities of the respondents need indeed to be questioned, as Cronbach's Alpha measuring the internal consistency of the scale which was designed to measure price sensitivities did not deliver acceptable results.

As the different datasets [(1)–(6)] were collected during different studies, the formulations of underlying questions might partly differ somewhat. Furthermore, the CNG/LPG adopters' evaluations of items concerning the derived three factors (cf. Table 5 in the Appendix) needed to be transformed from a 7-point scale to a 6-point scale in order to make them comparable with the evaluations in the other datasets. This might bias the results, too.

Amongst others additional limitations concerning representativity of the sample should be mentioned. Some of the subsamples are very small and strong self-selection effects motivating first movers to participate in the different studies are assumed.

# 4 Intentions of Fleet-EV Users to Privately Purchase an EV

## *4.1 Data Used*

In order to derive conclusions about BEV users' further adoption intentions characterized by their potential future purchase decision, the dataset of the EV company car users of the CROME project (the major part of the respondents in group (5), cf. Table 1) is further analyzed together with data originating from an earlier survey about these EV users' expectations that was distributed directly after the companies joined the CROME project. Joining the two datasets is possible due to the identifying user IDs that were attributed to the EV users at the beginning of the field test (cf. Ensslen et al. 2012, 2013a).

## *4.2 Methods Used*

First a comparison between the attitudes and norms of the BEV company car users potentially willing to privately purchase an EV and those users who are not willing or who are undecided to do so is conducted. Therefore t-Tests, nonparametric Mann-Whitney-Tests and a binary logistic regression analysis are applied. Furthermore, the highly significant dependencies between the factor *attitude towards EV* and the users' degree of satisfaction with different characteristics of the EV are analyzed and explained by applying linear regression analysis.

Additionally binary logistic regression analysis is performed in order to develop a model representing EV purchase intention by considering respondents' socio-demographic backgrounds, mobility patterns and their EV experience levels.

## 4.3 Results

### 4.3.1 Attitudes and Norms Influencing EV Purchase Intentions

Ensslen et al. (2013b) derived five factors potentially important for individuals' EV purchase intention by applying principal component analysis. The five factors under consideration are the individuals' *innovativeness, price sensitivity, environmental awareness, attitude towards EV* as well as the *perceived external image effect of EV*.

According to these findings intentions to privately purchase EV within the next years can neither be explained by respondents' environmental awareness nor by their price sensitivity (Fig. 4). According to t-Test results, respondents' innovativeness on the other hand discriminates at a marginally significant level between respondents who could envision purchasing an EV within the next years and those who cannot or are undecided. Furthermore, the factors *perceived external image effect of EV* and *attitude towards EV* discriminate between the two groups at a (highly) significant level. The users' *perceived external image effect of EV* was measured by four items. *Attitude towards EV* was also measured by four items. This factor is composed of items measuring EVs' relative advantage (i) (*I prefer driving an electric car to driving a conventional car*, and *The electric car excites me.*), their compatibility with personal attitudes, needs and experiences (ii) (*The electric car is useful in everyday life.*) as well as their simplicity of use (iii) (*Using the EV is easy.*). According to Rogers (2003) these characteristics of innovations are crucial for individuals' decisions to adopt or to reject an innovation. The factors *innovativeness* as well as *perceived external image effect of EV* further characterize EVs' compatibility (ii) with personal values as well as their compatibility with individuals' perceived social values and norms.

EVs' innovation characteristics trialability (iv) and observability (v) are pre-conditions within the CROME project, as all of the respondents had the possibility to drive the EV as the EV are part of their companies' fleets. Only at the beginning of the CROME field trial not all of the participants had already been using an EV (cf. Ensslen et al. 2013a), so importance of trialability can be analyzed. Detailed information about the methods and items used to measure (i)–(iii) as well as national factor scores for France and Germany can be found in Ensslen et al. (2013b).

In the beginning of the field trial participants were asked about their expectations concerning EV. As about 80 % of the respondents (cf. group (5), Table 1) had already experienced the EV when the survey about their expectations was distributed, additional data of a control sample of respondents interested in BEV was collected (cf. group (6), Table 1). Respondents were asked to which degree they think that different characteristics of EV will meet their expectations. The joined sample is divided into two subgroups. The subgroup which did not experience EV at all and the subgroup that experienced EV during at least one or two trips as driver or passenger. Results show that expectations of those who had not tested an EV

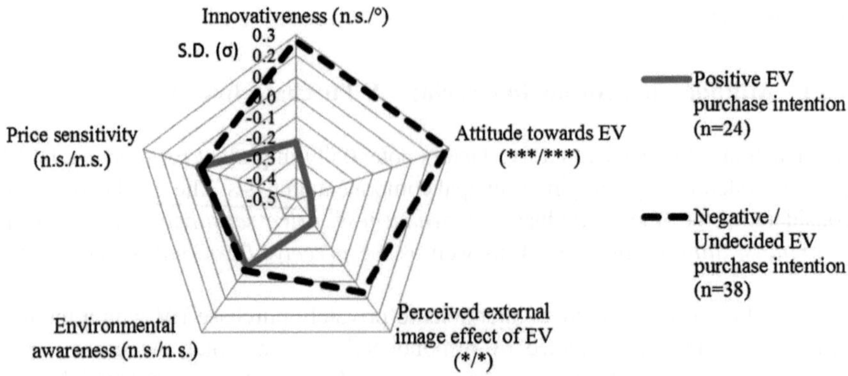


**Fig. 4** Arithmetic averages of PCA scores (second CROME survey, cf. Ensslen et al. 2013b) according to respondents' answers concerning their intention to purchase an EV within the next years (first CROME survey, cf. Ensslen et al. 2012) (Statistical Tests: (Mann-Whitney-Test / T-Test): (°/°): p < 0.1, (*/*): p < 0.05, (**/**): p < 0.01, (***/***): p < 0.001, (n.s./n.s.): not significant. Original scale: Items were measured on the following scale: 1: Strongly agree, 2: Agree, 3: Agree somewhat, 4: Rather disagree, 5: Disagree, 6: Strongly disagree)

before are lower, particularly concerning EVs' driving characteristics, safety and reliability aspects as well as aspects covering operating costs of EV (cf. Fig. 5). Respondents with some experiences with EV are for example more likely to evaluate the driving pleasure and the acceleration of EV better than respondents without any experience.

In order to explain the relations between EV users' attitudes and norms and their private EV purchase intention $P_a(y = 1)$ within the next years, binary logistic regression analysis with the three (marginally) significant factors attitude towards EV, perceived external image effect of EV and innovativeness is conducted (cf. Fig 4, Eqs. 1 and 2[4]). Detailed information about the way these dimensions were measured can be found in Ensslen et al. (2013b).

$$P_a(y = 1) = \frac{1}{1 + e^{-\beta_k}} \tag{1}$$

with

$$\beta_k = 0.786^* + 1.560x_{a1}^{**} + 0.684x_{a2} + 0.555x_{a3}^{\circ} + \epsilon \tag{2}$$

Description of the variables:

$y$ Dependent variable representing potential EV purchase intention within the next years (0: Negative or Undecided/1: Positive)

$x_{a1}$ Individuals' PCA score for *attitude towards EV*

[4]Significance level of Wald statistic: p < 0.1; *: p < 0.05; **: p < 0.01; ***: p < 0.001.

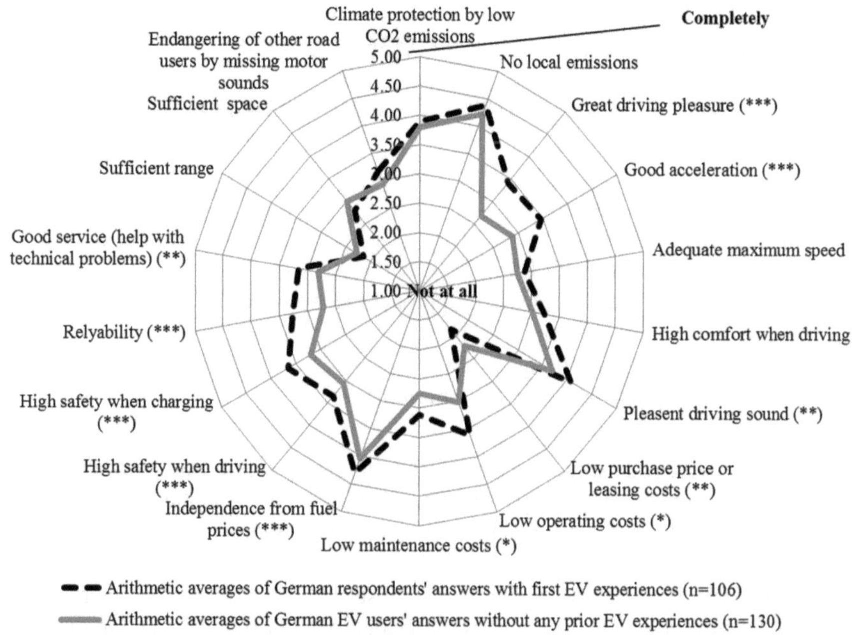


**Fig. 5** Arithmetic averages of respondents' evaluations about the degree to which they think that the EV will meet their expectations (Mann-Whitney-Test results: °: $p < 0.1$, *: $p < 0.05$, **: $p < 0.01$, *** : $p < 0.001$, n.s.: not significant. Original scale: 1: Not at all … 5: Completely.)

**Table 2** Quality criterions of the two binary logistic regression models $P_a$ and $P_c$ representing potential EV purchase intentions within the next years

| Model | n | Correctly classified (in %) | Nagelkerke $R^2$ | Cox and Snell $R^2$ | P-value of Hosmer and Lemeshow test |
|---|---|---|---|---|---|
| $P_a$ | 62 | 75.8 | 0.373 | 0.275 | 0.41 |
| $P_c$ | 180 | 68.9 | 0.296 | 0.221 | 0.131 |

$x_{a2}$ Individuals' PCA score for p*erceived external image effect of EV*
$x_{a3}$ Individuals' PCA score for i*nnovativeness*

$x_{a1}, x_{a2}$ and $x_{a3}$ are provided in standard deviations (σ). Details on quality criterions of this regression can be found in Table 2, details on this binary logistic regression in Table 6 in the Appendix.

As attitude towards EV ($x_{a1}$) highly impacts individuals' purchase intentions and high correlations between EV users' degree of satisfaction with different characteristics of EV and $x_{a1}$ could be observed (cf. Ensslen et al. 2013b), these dependencies are further analyzed. Therefore linear regression analysis is performed. The independent variables in the following equation can explain more than half of the variation ($n = 116$; $R^2 = 0.536$; Adjusted $R^2 = 0.506$) of $x_{a1}$.[5]

$$x_{a1} = -2.269^{***} + 0.891x_{b1}^{***} + 0.572x_{b2}^{***} + 0.357x_{b3}^{**} - 0.480x_{b4}^{**} + 0.326x_{b5}^{*} - 0.363x_{b6}^{*} + \epsilon \quad (3)$$

$x_{a1}$ Attitude towards EV ($\sigma$)
$x_{b1}$ Great driving pleasure
$x_{b2}$ General satisfaction with EV
$x_{b3}$ Safety of other road users when approaching noiseless
$x_{b4}$ High safety when driving
$x_{b5}$ High comfort when driving
$x_{b6}$ Climate protection by low $CO_2$ emissions

$x_{b1} - x_{b6}$ were measured on the following scale: (1) Not satisfied at all (2) Rather not satisfied (3) Rather satisfied (4) Completely satisfied

$x_{a1}$ represents PCA scores provided in standard deviations ($\sigma$).

Negative correlations between the attitude towards EV ($x_{a1}$) and the degrees of satisfaction with EVs' characteristics can be observed concerning climate protection by low $CO_2$ emissions ($x_{b6}$) as well as high safety when driving ($x_{b4}$). On the other hand $x_{a1}$ is positively correlated with individuals' degree of satisfaction concerning driving pleasure, the general satisfaction level with the EV, individuals' evaluations concerning safety of other road users when approaching noiseless and their indications concerning comfort level.

### 4.3.2 Explaining EV Purchase Intention with EV Users' Experience Levels, Income, Nationality and Mobility Needs

The following equation describes dependencies between individuals' intentions to purchase an EV within the next years and their mobility behaviors, their experience levels with EV, the number of cars in their households, their nationality, their income levels as well as their professional background as fleet manager (cf. Eqs 4 and 5[6]).

$$P_c(y = 1) = \frac{1}{1 + e^{-z_k}} \quad (4)$$

[5]Significance level of t-test: °: $p < 0.1$; *: $p < 0.05$; **: $p < 0.01$; ***: $p < 0.001$.

[6]Significance level of Wald statistic: °: $p < 0.1$; *: $p < 0.05$; **: $p < 0.01$; ***: $p < 0.001$. Please consider the reference categories provided in Table 5.

with

$$z_k = -1.567^* + 0.691x_{c1}^{\circ} + 2.079x_{c2}^* - 0.349x_{c3} - 0.970x_{c4}^* - 0.706x_{c5} + 0.600x_{c6} + 0.097x_{c7} + 2.062x_{c8}^{**} + 0.649x_{c9} + 0.587x_{c10}^* + \epsilon \quad (5)$$

Description of the variables:

- $y$ Dependent variable representing potential EV purchase intention within the next years (0: Negative or Undecided/1: Positive)
- $x_{c1}$ Travelled mileage on a (work)day (0: < 50 km/1: ≥ 50 km)
- $x_{c2}$ Fleet manager and user (0: No/1: Yes)
- $x_{c3}$ Respondent has experienced EV during one or two trips as a driver or passenger (0: No/1: Yes)
- $x_{c4}$ Respondent has not experienced EV so far at all (0: No/1: Yes)
- $x_{c5}$ Net household income <4,000 € (0: No/1: Yes)
- $x_{c6}$ Net household income ≥4,000 € (0: No/1: Yes)
- $x_{c7}$ Car usage frequency: 1–3 days per week (0: No/1: Yes)
- $x_{c8}$ Car usage frequency: 1–3 days per month or less (0: No/1: Yes)
- $x_{c9}$ French respondent (0: No/1: Yes)
- $x_{c10}$ Number of cars in the household (0-4 if $x_{c10} \leq 4$/5 if $x_{c10} > 4$)

Strong dependencies between the independent variable $x_{c4}$ describing whether the respondents have not experienced EV at all so far and the dependent variable EV purchase intention can be observed (the odds ratio Exp(B) is smallest for this variable). This should be further analyzed as this issue is supportive to the hypothesis that there might be wealthy districts where EV diffusion rates might be comparably higher as soon as some of the residents have adopted EV as innovators. This might be challenging the local distribution grids (Jochem et al. 2013 and Waraich et al. 2013).

### 4.3.3 Quality Criterions of the the Binary Logistic Regression Models

The quality criterions of both binary logistic regression models $P_a$ and $P_c$ are acceptable as values of Nagelkerke $R^2$ as well as of Cox and Snell $R^2$ serving as quality measure for the models, are at an acceptable level for both models (cf. Table 2, Backhaus et al. 2008). Furthermore, p-values for the Hosmer and Lemeshow Tests which analyzes the differences between the model results and the observed values are both not significant. Nevertheless, it needs to be addressed that uncertainties in all three models are high. Details of the models $P_a$ and $P_c$ are provided in Tables 5 and 6 in the Appendix.

### 4.4 Limitations

The independent variables $x_{a2}$ and $x_{c9}$ did not discriminate significantly between the respondents willing to privately purchase an EV within the next years and those not willing to do so. As these are the only variables in the models which are not significant ($p_{x_{a2}} = 0.12$ and $p_{x_{c9}} = 0.12$) and the samples are relatively small ($n_a = 62$ and $n_c = 180$), the authors assume that these results are due to the small samples considered. As the models would not be working without the variables it is important to take them into account, even if they are not significant. Furthermore, it needs to be mentioned that the results might be biased somewhat as the surveys were completed in French and German languages. Additionally, amongst others limitations concerning representativity of the sample should be mentioned. Particularly self-selection effects of organizations and EV users participating in the CROME field trial need to be considered when the results are interpreted.

## 5 Summary, Conclusions and Outlook

During the last years rather low market penetration rates for EV were observable. So far there are some first movers called innovators who privately purchased EV. Furthermore, there are some companies that purchased EV for their car pools. The users of these EV cannot be described as real innovators, as their companies made the decision to purchase the EV. Nevertheless, these persons are using and experiencing the EV technology. Although this group's innovativeness is comparably low, only one fifth of the respondents answered that they could not envision purchasing an EV within the next ten years (cf. Ensslen et al. 2012). These findings are supportive to the role of trialability for diffusion of innovations (Rogers 2003) and are supportive to Peters et al. (2011a, b) mentioning low-threshold possibilities to test EV in order to increase EV acceptance levels. Respondents who had the possibility to experience EV show comparably high levels of satisfaction with different characteristics of EV (cf. Fig. 5). This is further supported by analyzing who of the company EV users is rather willing to purchase an EV within the next years. According to the binary logistic regression model presented in this article, that was estimated based on survey data collected in the beginning of the French-German field operational test CROME, EV usage experience positively impacts EV purchase intentions. According to this model EV purchase intentions increase with a higher level of income, with a higher number of cars in the household and a daily mileage of more than 50 km not necessarily travelled by car. Furthermore, potential explanations for the fact that experiencing EV positively impacts EV purchase intentions are discussed. According to the results presented in this study EV purchase intentions can to a large part be explained by a factor representing the respondents' attitude towards EV (cf. Ensslen et al. 2013b). This factor can be explained by the users' degree of satisfaction with the EVs' driving characteristics, their satisfaction with EVs' safety characteristics as well as their satisfaction levels with EVs' ability to protect the climate by comparably low $CO_2$ emissions.

We have seen that early EV adopters differ from LPG and CNG users particularly concerning educational level, income and price sensitivity. In order to increase market shares of EV significantly and to reach early adopter stage (more than 2.5 % of cars are EV), EV sales activities should not only be targeted at individuals with low price sensitivities and high incomes. Furthermore, EV specific disadvantages compared to internal combustion engine vehicles, notably their comparably high prices and their limited range (cf. Fig. 5), give rise to the assumption that alternative business models targeting economic and range-specific aspects of EV could be supportive to early-stage EV diffusion. Considering that the battery is the most expensive part of an EV, already existing specific battery leasing options might reduce the initial high invest and so make EV affordable to a higher share of potential customers. As the economic perspective of an EV majorly depends on its vehicle miles travelled due to comparably lower variable costs (e.g. costs for electricity are lower than costs for fuel), business models like e.g. carsharing with EV could be a solution. First attempts to realize these business models can already be observed. However, whether they are going to be successful in the long run is unclear so far. Car sharing concepts might reduce the EV specific disadvantages, range and purchase price, from the users' perspectives significantly.

If policy makers wish to take measures in order to support private EV diffusion at the current market stage, thinking about demand side policy measures targeting monetary and non-monetary aspects could be an option. Non-monetary measures should besides the development of public accessible charging infrastructure including adequate parking space particularly focus on establishing possibilities to experience EV (e.g. test drives, e-car-sharing) as EVs' driving characteristics are exceeding the expectations. Possibilities to experience EV should particularly target opinion leaders within the early EV adopters, notably decision makers in organizations potentially being fleet managers. Furthermore, first stage marketing measures to potential private EV adopters should particularly target households with a high net income, equipped with two or more cars and having a high daily mileage. In order to further support possibilities to experience EV, policy makers could additionally think about providing incentives to make usage-oriented business models like e-car-sharing more attractive for potential operators.

Further analyses are planned in order to derive conclusions about EV users' decision concerning the powertrain choice during their next car purchase decision. Furthermore, the question is going to be addressed which services are favorable from the users' perspectives in order to compensate for BEV-specific barriers and how these should look like.

Furthermore, after first analyses have already been conducted, the binary logistic regression model explaining EV purchase intention with EV users' experience levels, income, nationality and mobility needs could be applied to representative mobility studies in France and Germany (MiD, ENTD) in order to equip every individual in these studies with EV purchasing probabilities. Exogenous variables which are not available in these studies are their experience levels with EV. However, it can be assumed that these are currently rather small. Furthermore, the research question can be expressed where EV diffusion will take place first. Will this be the case in rather urban or rather rural areas? The research question where early EV adoption will take place should be focused on, as local bottlenecks in the electric power grids due to electric mobility might occur.

**Acknowledgements** We would like to thank Verena Knödler and Mario Ostwald for their contributions and support concerning data collection and analyses. This research was made possible by the CROME project [ref. no. 01ME12002], the MeRegioMobil project [ref. no. 01ME9005] and the iZEUS project [ref. no. 01ME12013] funded by the German Federal Ministry of Economics and Technology (BMWi).

## Appendix A

See Tables 3, 4, 5 and 6.

**Table 3** Rotated component matrix

| Rotated Component Matrix[a] | Component | | | Communalities | Measures of sampling adequacy (MSA) |
|---|---|---|---|---|---|
| | 1[b] | 2[c] | 3[d] | | |
| It worries me when I think about the environmental conditions under which our children and grandchildren will probably have to live[e] | 0.86 | 0.02 | 0.01 | 0.75 | 0.63 |
| If we continue with business as usual, we are heading towards an environmental catastrophe[e] | 0.88 | −0.03 | 0.03 | 0.78 | 0.62 |
| The citizens can make significant contributions to climate protection by environmentally conscious everyday behavior[f] | 0.72 | 0.06 | 0.12 | 0.54 | 0.77 |
| I am very excited about technologies | 0.09 | 0.72 | 0.24 | 0.59 | 0.72 |
| I often seek out information about new products and brands[g] | 0.07 | 0.88 | 0.03 | 0.77 | 0.6 |

(continued)

**Table 3** (continued)

| Rotated Component Matrix[a] | | | | | |
|---|---|---|---|---|---|
| | Component | | | Communalities | Measures of sampling adequacy (MSA) |
| | 1[b] | 2[c] | 3[d] | | |
| I am often one of the first persons in my circle of friends and acquaintances, who is getting new technologies as soon as they appear on the market[h] | −0.10 | 0.82 | −0.18 | 0.71 | 0.59 |
| When I purchase products I compare them first and then buy the cheapest | 0.03 | −0.12 | 0.76 | 0.59 | 0.52 |
| When purchasing a product, I always try to maximize the quality I get for the money I spend[i] | 0.09 | 0.19 | 0.80 | 0.68 | 0.55 |
| Explained variance of factors | 25.92 | 25.15 | 16.49 | | |
| Cronbach's Alpha | 0.77 | 0.74 | 0.43 | | |
| Extraction Method: Principal Component Analysis. Rotation Method: Varimax with Kaiser normalization | | | | | |

[a]Rotation converged in 4 iterations
[b]Factor 1: Environmental awareness
[c]Factor 2: Innovativeness
[d]Factor 3: Price sensitivity
[e]cf. Kuckartz et al. (2006)
[f]cf. SINUS (2012)
[g]cf. Manning et al. (1995)
[h]cf. Parasuraman (2000)
[i]cf. Lichtenstein et al. (1993)

**Table 4** Adjusted significance levels concerning differences of the samples' levels of environmental sensitivity, innovativeness and price sensitivity

| Pairwise comparisons | | Adjusted significance levels | | |
|---|---|---|---|---|
| Sample 1 | Sample 2 | Environmental sensitivity | Innovativeness | Price sensitivity |
| E-scooter users | Persons interested in EV | n.s. | n.s. | * |
| E- scooter users | Persons interested in smart energy home technologies | n.s. | n.s. | ** |
| E- scooter users | Private EV customers | n.s. | n.s. | n.s. |
| E- scooter users | LPG and CNG users | n.s. | n.s. | ** |
| E- scooter users | EV company car users | ** | n.s. | n.s. |
| Persons interested in EV | Persons interested in smart energy home technologies | n.s. | *** | n.s. |
| Persons interested in EV | Private EV customers | n.s. | n.s. | ° |
| Persons interested in EV | LPG and CNG users | n.s. | *** | n.s. |
| Persons interested in EV | EV company car users | *** | n.s. | n.s. |
| Persons interested in smart energy home technologies | Private EV customers | n.s. | n.s. | * |
| Persons interested in smart energy home technologies | LPG and CNG users | n.s. | n.s. | n.s. |
| Persons interested in smart energy home technologies | EV company car users | *** | * | n.s. |
| Private EV customers | LPG and CNG users | n.s. | n.s. | * |
| Private EV customers | EV company car users | n.s. | n.s. | n.s. |
| LPG and CNG users | EV company car users | ° | ** | n.s. |

Each row tests the null hypothesis that the Sample 1 and the Sample 2 distributions are the same. Asymptotic significance (2-sided tests) are displayed
°: $p < 0.1$ | *: $p < 0.05$ | **: $p < 0.01$ | ***: $p < 0.001$ | n.s.: not significant

**Table 5** Variables in the equaton of model $P_c$

| | Regression coefficients B | Standard errors S.E. | Wald | df | Sig. | Exp (B) | 95 % C.I. for EXP (B) lower value | 95 % C.I. for EXP (B) upper value |
|---|---|---|---|---|---|---|---|---|
| $x_{c1}$ | 0.691 | 0.369 | 3.504 | 1 | 0.061 | 1.996 | 0.968 | 4.117 |
| $x_{c2}$ | 2.079 | 0.850 | 5.982 | 1 | 0.014 | 7.993 | 1.511 | 42.276 |
| | | | 4.709 | 2 | 0.095[a] | | | |
| $x_{c3}$ | −0.349 | 0.425 | 0.674 | 1 | 0.412 | 0.706 | 0.307 | 1.622 |
| $x_{c4}$ | −0.970 | 0.448 | 4.682 | 1 | 0.030 | 0.379 | 0.158 | 0.913 |
| | | | 8.701 | 2 | 0.013[b] | | | |
| $x_{c5}$ | −0.706 | 0.454 | 2.426 | 1 | 0.119 | 0.493 | 0.203 | 1.200 |
| $x_{c6}$ | 0.600 | 0.457 | 1.725 | 1 | 0.189 | 1.822 | 0.744 | 4.458 |
| | | | 7.244 | 2 | 0.027[c] | | | |
| $x_{c7}$ | 0.097 | 0.490 | 0.039 | 1 | 0.843 | 1.102 | 0.422 | 2.878 |
| $x_{c8}$ | 2.062 | 0.787 | 6.868 | 1 | 0.009 | 7.864 | 1.682 | 36.773 |
| $x_{c9}$ | 0.649 | 0.415 | 2.451 | 1 | 0.117 | 1.914 | 0.849 | 4.316 |
| $x_{c10}$ | 0.587 | 0.240 | 6.001 | 1 | 0.014 | 1.799 | 1.125 | 2.878 |
| Constant | −1.567 | 0.626 | 6.256 | 1 | 0.012 | 0.209 | | |

[a]Reference category for dichotomized variables $x_{c3}$ and $x_{c4}$: Respondent has experienced EV during several trips
[b]Reference category for dichotomized variables $x_{c5}$ and $x_{c6}$: Respondent did not want to provide information about the households' net income
[c]Reference category for dichotomized variables $x_{c7}$ and $x_{c8}$: Respondent (almost) daily using a car

**Table 6** Variables in the equaton of model $P_a$

| | Regression coefficients B | Standard errors S.E. | Wald | df | Sig. | Exp (B) | 95 % C.I. for EXP (B) lower value | 95 % C.I. for EXP (B) upper value |
|---|---|---|---|---|---|---|---|---|
| $x_{a1}$ | 1.560 | 0.506 | 9.504 | 1 | 0.002 | 4.759 | 1.765 | 12.831 |
| $x_{a2}$ | 0.684 | 0.437 | 2.444 | 1 | 0.118 | 1.981 | 0.841 | 4.669 |
| $x_{a3}$ | 0.555 | 0.328 | 2.868 | 1 | 0.090 | 1.742 | 0.916 | 3.313 |
| Constant | 0.786 | 0.355 | 4.907 | 1 | 0.027 | 2.195 | – | – |

# References


Backhaus K, Erichson B, Plinke W, Weiber R (2008) Multivariate Analysemethoden. Springer, Heidelberg

Bamberg S, Möser G (2007) Twenty years after Hines, Hungerford and Tomera: a new meta-analysis of psycho-social determinants of proenvironmental behaviour. J Environ Psychol 27:14–25

Bass FM (2004) Comments on a new product growth for model consumer durables the bass model. Manage Sci 50:1833–1840

BMVBS (Bundesministerium für Verkehr, Bau und Stadtentwicklung) (2011) Elektromoblilität—Deutschland als Leitmarkt und Leitanbieter. http://www.bmvi.de/SharedDocs/DE/Publikationen/VerkehrUndMobilitaet/elektromobilitaet-deutschland-als-leitmarkt-und-leitanbieter.pdf?__blob=publicationFile. Accessed 6 March 2014

Bundesnetzagentur (2011) „Smart Grid" und „Smart Market": Eckpunktepapier der Bundesnetzagentur zu den Aspekten des sich verändernden Energieversorgungssystems. http://www.bundesnetzagentur.de/SharedDocs/Downloads/DE/Sachgebiete/Energie/Unternehmen_Institutionen/NetzzugangUndMesswesen/SmartGridEckpunktepapier/SmartGridPapierpdf.pdf?__blob=publicationFile&v=2%20. Accessed 6 March 2014

Bundesregierung (2009) Nationaler Entwicklungsplan Elektromobilität der Bundesregierung. http://www.bmbf.de/pubRD/nationaler_entwicklungsplan_elektromobilitaet.pdf. Accessed 6 March 2014

Dütschke E, Schneider U, Peters A, Paetz A, Jochem P (2011) Moving towards more efficient car use—what can be learnt about consumer acceptance from analysing the cases of LPG and CNG. ECEEE 2011 Summer Study

Dütschke E, Schneider U, Peters A (2013) Who will use electric vehicles?. Working paper sustainability and innovation No. S 6/2013. Fraunhofer Institute for Systems and Innovation Research (ISI). http://www.isi.fraunhofer.de/isi-media/docs/e-x/working-papers-sustainability-and-innovation/WP06-2013_Electric_Vehicles.pdf. Accessed 7 March 2014

Dunn OJ (1964) Multiple comparisons using rank sums. Technometrics 6:241–252

Eckl-Dorna W, Sorge N-V (2013) Elektroauto-Markt wächst exponentiell, nicht linear. Interview with Kagermann, Henning. Manager magazin online 2013. http://www.manager-magazin.de/unternehmen/autoindustrie/a-901804.html. Accessed 12 July 2013

Ensslen A, Babrowski S, Jochem P, Fichtner W (2012) Existe-il des différences d'acceptation des véhicules électriques entre la France et l'Allemagne? – Premiers résultats de l'analyse scientifique du test de flotte Cross Border Mobility for Electric Vehicles (CROME). Proceedings of the 11ème séminaire francophone est-ouest de socio-économie des transports, Karlsruhe

Ensslen A, Jochem P, Schäuble J, Babrowski S, Fichtner W (2013a) User acceptance of electric vehicles in the French-German transnational context, in selected proceedings of the 13th WCTR, Rio de Janeiro

Ensslen A, Jochem P, Fichtner W (2013b) Experiences of EV users in the French-German context. In: Proceedings of EVS27-conference, Barcelona, Spain

Eurostat (2013a) Umwelt und Energie Datenbank. http://epp.eurostat.ec.europa.eu/portal/page/portal/climate_change/data/main_tables. Accessed 18 Sept 2013

Eurostat (2013b) Energy price statistics. http://epp.eurostat.ec.europa.eu/statistics_explained/index.php/Energy_price_statistics. Accessed 25 June 2013

European Commission (2011) A roadmap for moving to a competitive low carbon economy in 2050. http://ec.europa.eu/clima/policies/roadmap/documentation_en.htm. Accessed 3 March 2014

Fraunhofer Institute for Systems and Innovation Research (ISI) (2012) REM2030 Driving Profiles Database V2012

Frenzel I, Jarass J, Trommer S, Lenz B (2015) Erstnutzer von Elektrofahrzeugen in Deutschland. Nutzerprofile, Anschaffung, Fahrzeugnutzung. Berlin: Deutsches Zentrum für Luft- und Raumfahrt e. V. (DLR)

Glerum A, Stankovikj L, Thémans M, Bierlaire M (2013) Forecasting the demand for electric vehicles: accounting for attitudes and perceptions. Transportation Science, Articles in Advance, pp 1–17

Hackbarth A, Madlener R (2013) Consumer preferences for alternative fuel vehicles: a discrete choice analysis. Transp Res Part D Transp Environ 25:5–17

Hartung J, Elpelt B, Klösener K-H (2005) Statistik. Oldenbourg, München

IBM SPSS Statistics (2013a) Kruskal-Wallis one-way analysis of variance (nonparametric tests algorithms). http://pic.dhe.ibm.com/infocenter/spssstat/v21r0m0/index.jsp?topic=%2Fcom.ibm.spss.statistics.help%2Falg_npar_tests_kruskalwallis.htm. Accessed 18 Sept 2013

IBM SPSS Statistics (2013b) Beispiel: Boxplot. http://pic.dhe.ibm.com/infocenter/spssstat/v20r0m0/index.jsp?topic=%2Fcom.ibm.spss.statistics.help%2Fgraphboard_creating_examples_boxplot.htm. Accessed 18 Sept 2013

IBM SPSS Statistics (2012) Post hoc comparisons for the Kruskal-Wallis test. http://www-01.ibm.com/support/docview.wss?uid=swg21477370. Accessed 22 Oct 2013

Jochem P, Kaschub T, Fichtner W (2013) How to integrate electric vehicles in the future energy system? In: Hülsmann M, Fornahl D (eds) Evolutionary paths towards the mobility patterns of the future. Springer, Heidelberg

KBA—Kraftfahrt-Bundesamt (2014) Jahresbilanz des Fahrzeugbestandes am 1. Januar 2014. http://www.kba.de/cln_031/nn_125398/DE/Statistik/Fahrzeuge/Bestand/2014__b__jahresbilanz.html#rechts. Accessed 19 March 2014

Kay D, Hill N, Newman D (2013) Powering Ahead—The future of low-carbon cars and fuels. http://www.theengineer.co.uk/Journals/2013/04/22/i/k/h/powering_ahead-kay_et_al-apr2013-embargoed_copy.pdf. Accessed 7 March 2014

Kuckartz U, Rädiker S, Rheingans-Heintze A (2006) Umweltbewusstsein in Deutschland. Marburg. http://www.umweltbundesamt.de/publikationen/umweltbewusstsein-in-deutschland-2006. Accessed 10 March 2014

Lichtenstein D, Ridgway N, Netemeyer R (1993) Price perception scales. In: Bearden W, Netemeyer R, Haws K (2011) Handbook of marketing scales. SAGE Publications, Thousand Oaks

Mabit SL, Fosgerau M (2011) Demand for alternative-fuel vehicles when registration taxes are high. Trans Res D 16:225–231

Manning K, Bearden W, Madden T (1995) Innovativeness: consumer innovativeness. In: Bearden W, Netemeyer R, Haws K (2011) Handbook of marketing scales. SAGE Publications, Thousand Oaks

Mendes Lopes M, Moura F, Martinez LM (2014) A rule-based approach for determining the plausible universe of electric vehicle buyers in the lisbon metropolitan area. Transp Res Part A 59:22–36

MOP (2012) Mobilitätspanel Deutschland (MOP). http://www.clearingstelle-verkehr.de. Accedded 7 March 2013

Paetz AG, Dütschke E (2012) Auf dem Weg in die elektromobile Zukunft - Ein Zwischenfazit zur Elektromobilität im Alltag. SIV. news 1:45–47

Paetz A-G, Kaschub T, Jochem P, Fichtner W (2012a) Demand response with smart homes and electric scooters: an experimental study on user acceptance. http://www.aceee.org/files/proceedings/2012/data/papers/0193-000232.pdf. Accessed 7 March 2014

Paetz A-G, Jochem P, Fichtner W (2012b) Demand Side Management mit Elektrofahrzeugen – Ausgestaltungsmöglichkeiten und Kundenakzeptanz. 12th Symposium Energieinnovation. Graz, Austria. http://portal.tugraz.at/portal/page/portal/Files/i4340/eninnov2012/files/lf/LF_Paetz.pdf. Accessed 7 March 2014

Paetz A-G, Kaschub T, Pfriem M, Jochem P, Fichtner W, Gauterin F (2013) Is electric mobility a means for more sustainability?. In: Behavior, energy and climate change conference, Nov 2013, Sacramento

Parasuraman A (2000) The technology readiness index. In: Bearden W, Netemeyer R, Haws K (2011) Handbook of Marketing Scales. SAGE Publications, Thousand Oaks

Peters A, Dütschke E (2010) Zur Nutzerakzeptanz von Elektromobilität. Analyse aus Expertensicht. Fraunhofer Institute for Systems and Innovation Research (ISI). http://publica.fraunhofer.de/eprints/urn:nbn:de:0011-n-1450132.pdf. Accessed 7 March 2014

Peters A, Hoffmann J (2011) Nutzerakzeptanz von Elektromobilität. Eine empirische Studie zu attraktiven Nutzungsvarianten, Fahrzeugkonzepten und Geschäftsmodellen aus Sicht potenzieller Nutzer. Fraunhofer Institute for Systems and Innovation Research (ISI). http://isi.fraunhofer.de/isi-media/docs/service/de/presseinfos/Forschungsergebnisse_Nutzerakzeptanz_Elektromobilitaet.pdf. Accessed 7 March 2014

Peters A, Popp M, Agosti R, Ryf B (2011a) Electric mobility—a survey of different consumer groups in Germany with regard to adoption. ECEEE Summer Study proceedings. http://publica.fraunhofer.de/eprints/urn:nbn:de:0011-n-1834806.pdf. Accessed 7 March 2014

Peters A, Gutscher H, Scholz RW (2011b) Psychological determinants of fuel consumption of purchased new cars. Transp Res Part F Psychol Behav 14:229–239

Peters A, de Haan P, Scholz W (2013) Understanding car buying behavior: psychological determinants of energy efficiency and practical implications. Int J Sustain Transp doi:10.1080/15568318.2012.732672

Pfahl S (2013) When will electric vehicles capture the German market? And why? In: Proceedings of EVS27-conference, Barcelona, Spain

Plötz P, Schneider U, Globisch J, Dütschke E (2014) Who will buy electric vehicles? Identifying early adopters in Germany. Transportation Research Part A: Policy and Practice 67:96–109, doi:10.1016/j.tra.2014.06.006

Rogers EM (2003) Diffusion of innovations, 5th edn. Free Press, New York

SINUS (2012) Klimaschutz in Heidelberg. http://www.uni-heidelberg.de/md/journal/2013/02/14jan_heidelbergstudie_2012.pdf. Accessed 7 March 2014

Thielmann A, Sauer A, Isenmann R, Wietschel M, Plötz P (2012) Produkt-Roadmap Lithium-Ionen-Batterien 2030. Fraunhofer Institute for Systems and Innovation Research (ISI). http://www.isi.fraunhofer.de/isi-media/docs/v/de/publikationen/PRM-LIB2030.pdf. Accessed 7 March 2014

Waraich RA, Galus MD, Dobler C, Balmer M, Andersson G, Axhausen KW (2013) Plug-in hybrid electric vehicles and smart grids: investigations based on a microsimulation. Transp Res Part C 28:74–86

Wietschel M, Dütschke E, Funke S, Peters A, Plötz P, Schneider U (2012) Kaufpotenzial für Elektrofahrzeuge bei sogenannten „Early Adoptern". Fraunhofer Institute for Systems and Innovation Research (ISI). http://isi.fraunhofer.de/isi-media/docs/e/de/publikationen/Schlussbericht_Early_Adopter.pdf. Accessed 7 March 2014

Wietschel M, Plötz P, Kühn A, Gnann T (2013) Markthochlaufszenarien für Elektrofahrzeuge. Fraunhofer Institute for Systems and Innovation Research (ISI). http://www.isi.fraunhofer.de/isi-media/docs/e/de/publikationen/Fraunhofer-ISI-Markthochlaufszenarien-Elektrofahrzeuge-Langfassung.pdf. Accessed 7 March 2014

Windisch E (2013) Driving electric? A financial assessment of electric vehicle policies in France. Paris

Ziegler A (2012) Individual characteristics and stated preferences for alternative energy sources and propulsion technologies in vehicles: a discrete choice analysis for Germany. Transp Res A 46:1372–1385

## Author Biographies

**Axel Ensslen** is research associate at the French-German Institute for Environmental Research and member of the research group Transport and Energy at the Chair of Energy Economics since 2011. He studied industrial engineering at the Karlsruhe Institute of Technology (KIT), Germany and the Ecole Nationale Supérieure des Mines de Nancy (ENSMN), France. His research activities focus on socio-economic aspects of electric mobility.

**Alexandra-Gwyn Paetz** is a Ph.D.-candidate at the Chair of Energy Economics at KIT. Her main research activities focus on new technologies in the electricity sector, such as smart grids, smart homes and electric vehicles. In 2007 she graduated in business economics at the Ludwig-Maximilians-University Munich. Today she works as executive assistant to the President of KIT.

**Sonja Babrowski** works since 2011 as a research associate at the Chair of Energy Economics of the Institute for Industrial Production (IIP) of KIT. In 2010 she graduated in industrial engineering at KIT. Her main area of research is energy system modeling with a focus on electric vehicles and electricity storage systems.

**Patrick Jochem** is since 2009 leader of the research group Transport and Energy at the Chair of Energy Economics at KIT. Since 2012 he is senior researcher at the Karlsruhe Service Research Institute (KSRI). In 2009 he received his Ph.D. in transport economics from Karlsruhe University. He studied economics at the universities in Bayreuth, Mannheim and Heidelberg, Germany. His research interests are in the fields of electric mobility and ecological economics.

**Wolf Fichtner** is Director of the Institute for Industrial Production and the French-German Institute for Environmental Research. He is professor and holder of the Chair of Energy Economics at KIT. His main areas of research are Energy System Modelling and the Techno-economic Analysis of Energy Technologies.

# How to Increase the Attractiveness of Electric Cars as Fleet Cars?

Jasmin Nehls

**Abstract** Both in public and scientific discussion, electric cars are claimed to suit as company cars (Rother and Eisert 2013). Particularly larger fleets may allow for the application of diversified carriers and within others of short distance vehicles. Here electric cars seem to fit. Due to increasing pressure on companies to save carbon dioxide, interest in alternative drive technologies has increased. **Recent surveys however indicate that companies particularly base acquisition decisions on budget and image criteria** (Holzer 2012). Since the electric car is neither known to be cheap, even from a Total Cost of Ownership (TCO) point of view (Nationale Plattform Elektromobilität 2011), nor particularly fancy (exceptions like Tesla's famous sports cars aside) initial interest in the technology in many cases doesn't materialize. The paper therefore explores ways to increase the attractiveness of electric cars. **Services provide benefits; their targeted design in order to increase economic payoff is however crucial**. The paper therefore first analyzes promising applications of the electric car within company fleets, which are then aggregated to five key purchase motives. It then elaborates service concepts tackling image and TCO issues of the electric car. Discussion of added value reveals potential for the successful integration of electric cars into corporate fleets and includes several best practice examples.

**Keywords** Electric mobility · Corporate fleet · Service

## 1 Introduction

The market development of electric cars reveals an ambiguous situation. On the one hand, the estimated uptake did not yet take place, considering that only 7,114 battery electric and 64,995 hybrid electric vehicles had been registered in Germany

J. Nehls (✉)
Institute for Integrated Product Development at the University of Bremen,
Badgasteiner Straße 1, 28359 Bremen, Germany
e-mail: jas@biba.uni-bremen.de



D. Fornahl and M. Hülsmann (eds.), *Markets and Policy Measures in the Evolution of Electric Mobility*, Lecture Notes in Mobility,
DOI 10.1007/978-3-319-24229-3_4

until 01/2013 (Kraftfahrt-Bundesamt 2013). On the other hand, improvement has occurred with regard to offers: The variety of available models has increased. Besides, new players, traditional German car manufacturers, such as Daimler, VW and BMW, have entered or were about to enter the market. They offer electrified versions of popular models such as the Golf or smart.

The key target of 2013's activities was to conquer the mass market: most models announced belonged to the compact or medium class. Financing opportunities became more flexible and adaptable to customer needs. Since electric cars are initially more expensive than conventional ones—by around 10,000 €—battery leasing offers a way to spread the cost over the period of use, therefore lowering upfront investments.

Furthermore, complementing services addressing maintenance and charging or a mobility guarantee were included in customized service packages made available through a growing network of car dealerships and service partners. Customers benefit from convenient access to these offers, particularly because they are already part of the leasing contract. Besides, many offers were made available online and their costs, though depending on individual needs, became increasingly transparent thanks to tariff calculators.

Besides addressing private users, the mentioned developments do strengthen the competitive position of electric cars for corporate fleets, particularly in case of large companies. These keep track of market development of electric cars but so far rarely purchase the vehicles available for a variety of reasons—with image and cost consideration playing the largest role:

On the one hand, **image** is a crucial factor: the larger the company, the greater the concern articulated within the car policy: In many cases, German companies stick to German quality car brands, such as VW, Mercedes, Audi or BMW—which so far barely offer electric cars (manager magazin online 2013).

There is, however, a recent emergence of a "green car policy" which integrates emission reductions—particularly in the field of greenhouse gases—as target in acquisition strategies (Telekom Mobility Solutions Webpage 2010):

Companies also use environmentally friendly technologies as part of their Corporate Social Responsibility strategy (CSR)—or more generally—for marketing purposes. Besides motivation to increase public awareness, policy pressure exists since the German government itself is bound by emission reduction targets set forth in international agreements (Hannoversche Allgemeine 2013).

A couple of initiatives providing the exchange of best practice, such as "Wirtschaft pro Klima" (Economy in favor of climate) (Wirtschaft pro Klima 2013), are funded by the Federal Ministry for the Environment, Nature Conservation and Nuclear Safety. Environmentally friendly transportation is one of the cornerstones and several companies present projects centering on electric cars within this domain.

Opportunities for generating a **positive image** therefore exist with the electric car, which may also bear commercial potential. These effects are yet difficult to estimate and case depending.

**The Cost** of electric cars stands as an important barrier and is compounded by other competitive disadvantages, such as short range, long charging times and limited availability of supporting infrastructure.

There is little potential for compensation, i.e. the reduction of operating costs owing to the comparatively lower price of electricity vis-à-vis fossil fuels. Over the medium-term, relative price developments of fossil fuel and electricity are hard to predict. Whether or not it will be possible to stabilize oil production worldwide, exploit largely untapped resources, such as bituminous sands, or increase efficiency of drive technology will have effects on the oil price. Simultaneously, investments necessary for ensuring continued long-term growth of renewable energy generation, which relies as much on generation as on transmission capacities for electrical energy, determine operation cost as well as sustainability of electric cars.

The market development observed so far does not provide enough security to allow a reliance on—let alone quantification of—lower operation costs of the electric car, therefore rather putting it at economic disadvantage also from total cost of ownership point of view.

When considering service offerings which can strengthen the cost effectiveness of the electric car, one needs to closely relate to the strengths of the technology. Therefore, the following paragraph will consider to which businesses electric cars can prove most beneficial.

## 2 Challenges and Opportunities of Electric Cars

Company fleet management is affected by overall economic development, (social) trends and the corporate strategy. Table 1 summarizes key developments of relevance to a possible inclusion of electric cars in company fleets. The overview is meant to sharpen the understanding of the status quo.

Particularly challenging for battery electric cars is the higher procurement cost. Furthermore, an increasing number of fleet managers' duties throughout the vehicle lifecycle, such as providing additional flexibility in employee mobility, call for complexity reduction. The electric car as "new" technology which needs particular planning, infrastructure solutions and employee communication is less attractive in this context. However, the electric car has some unique properties, which make it suitable for various business applications—Table 2 contains a short summary.

There are several technical properties such as **noise reduced driving** allowing for new fields of application, i.e. as night carriers, or catering to particular driving behavior. Emission reduction and avoidance (if combined with renewable energies), allows the electric car to be part of a **sustainable transportation concept**. These positive environmentally friendly properties of the electric car can serve a particular

**Table 1** Developments affecting company fleet management and their meaning to electric car integration

| Development | Rather positive effects on integration of electric cars | Rather negative effects on integration of electric cars |
|---|---|---|
| Increasing pressure to reduce fleet cost (Holzer 2012) | | **X** |
| Acquisition based on TCO considerations (Holzer 2012) | X (electric cars rarely "win" TCO calculations) | |
| Executive board pays little attention to fleet management (Wirtschaft pro Klima 2013) | | X (clear commitment to e-mobility by executive board necessary for introduction) |
| Increased complexity (BME 2000) | x (environmental requirements; changing mobility needs) | x (need for simplification) |
| Efficient fleet management (Vivaldini 2012) | | x (diversity increases complexity) |
| Partial outsourcing of fleet management (Vivaldini 2012) | x (if proficiency allows for holistic consideration of vehicles) | |
| Trend to drive large business cars, SUV (Fuhrpark+management 2012) | | x |
| Reconcile environmental performance, reliability, cost and comfort (Corporate Vehicle Observatory 2012) | | X (yet hard to provide economic benefits) |

green image going beyond reliability or technical perfection majorly addressed by car brand image in general. In the following, this distinction is captured under the **"Ecology"** motive. Particularly when leading to **new business concepts** or if triggered by legal restrictions pertaining to noise or exhaust emissions, economic benefits can be increased. Lower operating cost enhanced by added value improves TCO balance of the electric car.

Last but not least, the "alternative technology", which yet imposes challenges to the user, may serve as a **door opener to research** in car design, construction materials, battery development, energy systems and charging technology.

The general approach followed by traditional and new market players can be titled "smart": new vehicle shapes, such as BMW i3's cubic form, light materials, such as carbon fiber, battery packs with particular service packages comprising functional guarantee or maintenance as well as roaming in different electric grids. These may contribute to an **innovative image** of electric cars—and attract companies.

**Table 2** Promising characteristics of the electric car

| Exclusive property | Promising application | Type of vehicle | Social trends reinforcing the advantage of electric cars | Fure relevance |
|---|---|---|---|---|
| Noiseless driving, particularly at speeds below 50 km/h (Umweltbundesamt 2013) | Goods transportation at night, municipal services | Utility vehicles | Increasing number of home deliveries; request for locally produced goods and short distance transportation, noise reduction policy | Increasing relevance, needs legislative enforcement |
| Renewable sources of energy | Sustainable transportation; increased credibility of environmental goals within CSR | Cars, utility vehicles | Increasing public awareness and number of regulations regarding $CO_2$ emission reduction | Increasing attention, often not yet decisive advantage |
| Avoidance of local emissions | Improvement of air quality in urban areas; protection of building shell, $CO_2$ emission reduction | All | International regulations on emission reductions and industrial limitations | Increasing importance, first green zones in cities not yet challenging conventionally fuelled cars in general |
| Electric drive technology | Rapid acceleration | All | | Positive, needs to be experienced |
| New vehicle design concepts | Placement of the battery, interior space concepts | All | Trend towards customized goods | Increasing relevance |
| Use of latest technologies and materials typically found only in luxury cars | Chassis (partly) made from carbon fibre | All | Increasing international demand for premium products | High relevance |
| Innovative solutions in general | Heating with bio-ethanol | All | | Positive when range extending, rarely demand decisive |
| | Sustainability throughout vehicle lifecycle | | | |

## 3 Purchase Criteria

When summing-up key-findings on challenges and opportunities discussed above, five major criteria affect purchase decisions:

- Image
- Total cost of ownership
- Ecology
- Innovation: development of new products and services
- Facilitation of daily business

"Image" is pointing to reliability/quality product therefore not comprising green properties, which are referred to separately as "Ecology".

Since new business models generated around electric mobility such as consulting, fleet management, smart home design or others do not fit into the TCO balance of the electric car application itself, it is separately referred to as "new products and services". The same goes for the facilitation of business, i.e. by obtaining access to environmental zones. Though directly referring to the use of electric cars, the added value is hard to estimate in monetary terms.

Figure 1 depicts the interrelation between these five criteria in the context of their importance to particular interest groups.

Particularly when business is closely related to ecology and sustainability **credibility and competency** are enhanced and demonstrated to customers through the use of electric cars. **Electricity generating companies, the municipal services sector, enterprises offering transportation services and consultancies** are the focus.

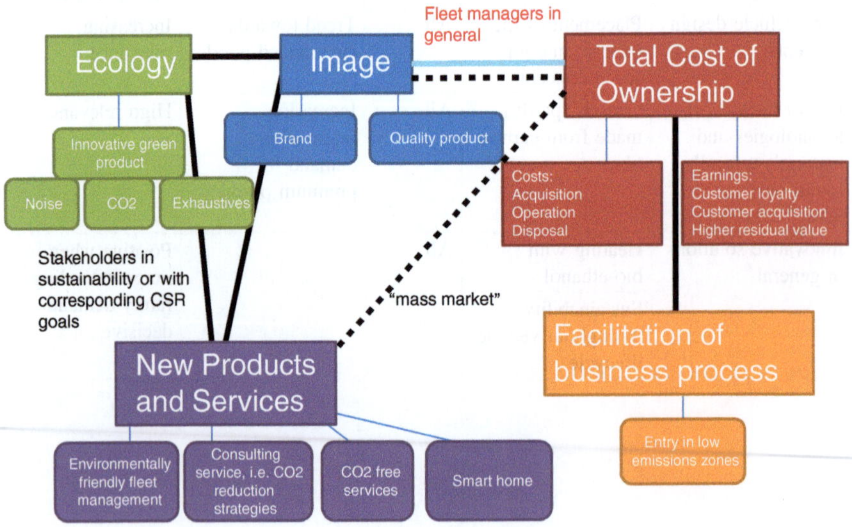


**Fig. 1** Relevant criteria positively influencing electric car application in company fleets

Companies operating in these domains may use experience with the electric car in order to **design new products and/or services** or complement and improve existing ones. Electricity suppliers which have developed new energy management and smart housing services offer a new "eco-mobility" service package to their customers.

Furthermore, **(large) companies** committed to **environmental protection** as addressed in their **CSR policy,** may use electric cars and other alternative drive-technologies to realize their goals. One such example is Telekom Mobility Solutions which has been implementing a green car policy since 2010 targeting an average $CO_2$ emission of 110 g/km for new company cars (Telekom Mobility Solutions Webpage 2010).

In order to appeal to businesses which cannot create a direct link between own activities and green transportation TCO benefits are crucial. The electric car may allow the **improved handling of daily routine task** for both companies and their customers. Access to low emission zones represents just one situation in which electric cars outperform standard vehicles, particularly older ones. The advantage may reach monetary level which is however hard to estimate. However there is a connection to TCO.

As mentioned, electric cars are not yet fully addressing the five purchase criteria—which would increase attractiveness to both to the mass market and green business sector. This can be seen from the slow market development. The electric car is by itself neither environmentally friendly nor cheap. This is where service comes into play.

In the following, services offerings with particular impact on TCO will be described and/or developed, because of the relevance of this criterion.

## 4 Service Increasing Economic Potentials

TCO consideration from a company's perspective comprises all the costs arising to a company over the lifecycle of a particular product and/or service. Correspondingly, there are three options to tackle reduction of TCO: at the beginning of life (product/service definition and planning), mid of life (use phase) and end of life (re-use/recycling phase).

Companies purchasing vehicles often are not capable to conduct whole lifecycle assessments. However, particularly due to increased volatility and insecurity in supply of fossil fuels, the mid of life behavior is gaining more interest by companies (Krischun 2010). This provides consultancies and fleet operators with the potentials of new consultancy service and i.e. fleet assessments. Taking a look at the possibilities for reducing TCO of electric cars, Fig. 2 provides an overview on core possibilities:

Besides the MoL phase, individual solutions within BoL and EoL phase bear further cost reduction potentials:

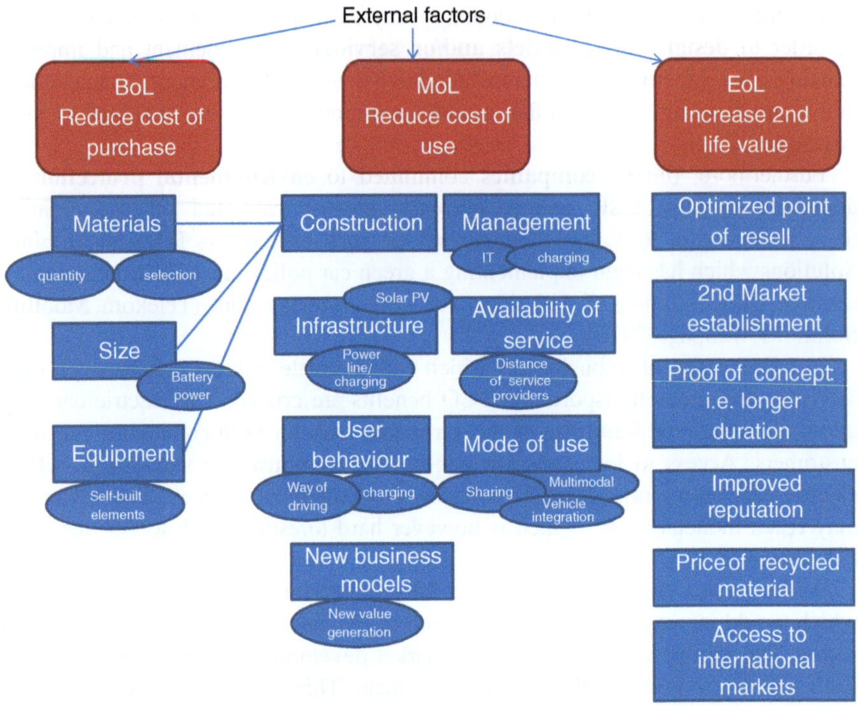


**Fig. 2** Possibilities for reduction of TCO of electric cars throughout product lifecycle

At the beginning of life, consideration of the particular mobility requirements in order to influence design of the vehicle allows for a new level of purpose design—very much individualized to the need of the company. Currently such a service is hardly available—Opel offers individual design of vehicles to customers (Opel Webpage 2013), while having a standardized construction pattern. Highly individualized service to electric mobility is in many cases provided by small to very small or start-up companies. The StreetScooter GmbH (StreetScooter GmbH; Website http://www.streetscooter.eu/fleet-customization.html) e.g., a spin-off of the technical University of Aachen and 20 medium-sized businesses, offers fleet customization and claims **TCO advantages** of its products. The company has furthermore developed a city car with variable maximum range of 40/80/120 km selectable by the customer according to the needs. Among big car manufacturers, the approach is closest followed by Renault, having developed the Twizy quad with 4 and 13 kW electric power (Renault webpage: http://www.renault.de/renault-modellpalette/ze-elektrofahrzeuge/twizy/twizy/einfuehrung/). However, Renault does not offer individualized manufacturing.

As a matter of fact, once the price of the electric vehicle is similar to a conventionally fuelled one and both are satisfying everyday mobility needs of the company (in terms of range, trunk space, speed, comfort etc.), TCO due to lower price of electricity and average maintenance expenditure may fall in favor of the electric car [see i.e. Kampker et al. (2013) for reference]. Research on these particular business cases is closely linked to the mid-of-life requirements and conditions of the company in question. Services, particularly in consulting and training can help shape a favorable **infrastructure** comprising

- parking and charging opportunities for electric vehicles,
- solar PV,
- availability of O&M service partners close-by,
- management software to facilitate access, use and data processing of the vehicles

in order to continuously improve vehicle integration into the company fleet. As a matter of fact, companies producing electricity from solar and wind closely consider electric mobility these days in search for possible **new business models** (see i.e. Sun et al 2009; Meschtscherjakov 2009).

**A favorable handling** also facilitates economics of application of electric cars, offering room for service such as information, inclusion of employees in tailored solution development or drive trainings, tackling

- Acceptance
- Way of driving
- Way of charging
- Organization of a vehicle pool

These factors can become decisive for suitability of electric cars by avoiding unnecessary damage or restrictions to mobility.

Besides, when addressing new business models, there is huge room for services at the intersection company–car–driver–environment. As the feasibility of the electric car is being pushed forward not only on a product but majorly on a service level, Fig. 3 depicts a vision:

Within Fig. 3 ICT services based on data processing from company, car, driver and environment are being developed and directly offered through the electric car. One exemplary application can occur in case of traffic congestion: since a company's schedule is tight and driver's priorities do not allow to skip reaching destinations planned (i.e. because a delivery must be made) a rerouting service based on up-to date information from other drivers of the car-user community may save time, increase comfort to the driver and improve the delivery service to the customer. Increased company reputation and eventually increased turnover may result. Another service going a step beyond route optimization comprises ad hoc service being offered by other members of the professional network of the company: a business partner caught in the same traffic jam can offer to take over the delivery to a common customer therefore saving waiting time and helping reduce the amount of cars within the jam.

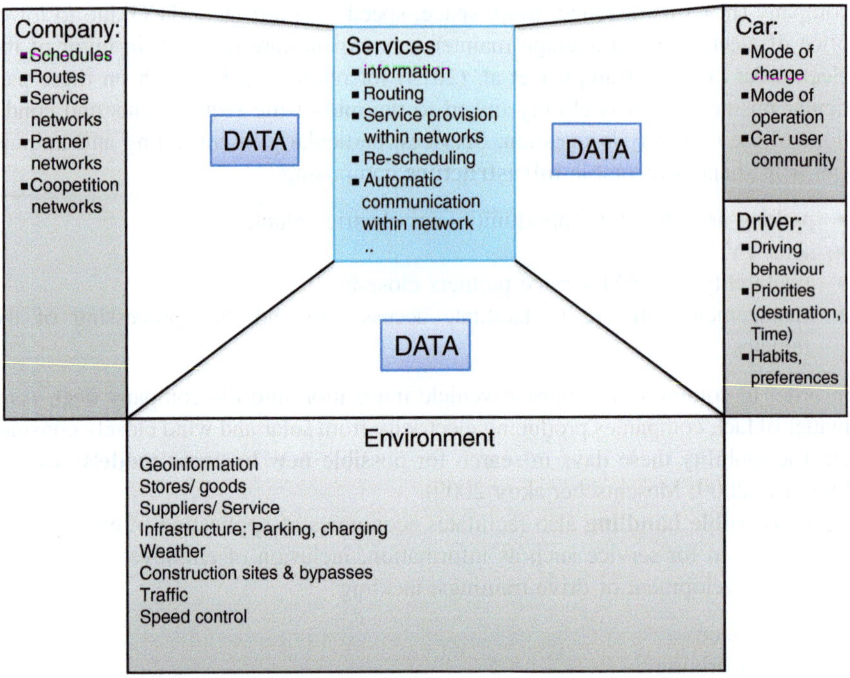


**Fig. 3** Services on data integration

Crucial determinants for such highly sophisticated ICT service comprise

- Identification and Trust—creation of ad hoc networks
- Openness
- Data safety
- Accountability

Relevant research on smart cars covers interaction of driver to car, car to environment as well as car to car (see i.e. Sun et al. 2009; Meschtscherjakov 2009). Major research questions are on predictability of user behavior with regard to the context, safety, comfort and user acceptance. Hence focus is usually on the driver—but not on the purpose of the drive. However, when drives are conducted only in the context of business, facilitation needs to take particular requirements of the business into account. If cars do actively improve business, companies gain added value within MoL phase.

In principle these services can be incorporated in any type of car. However, they can well be integrated into the three pillars "Economy", "Ecology" and "People" describing sustainable products such as environmentally friendly cars. Figure 4 highlights the idea:

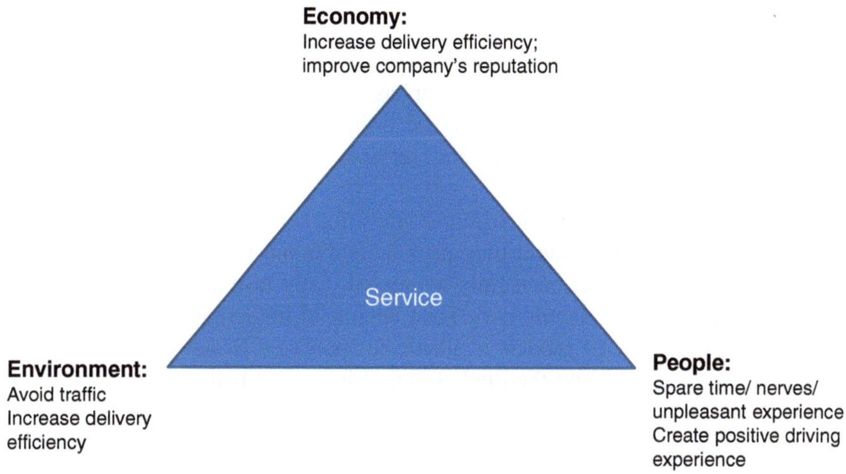


**Fig. 4** Three dimensions of sustainability based on Lexikon der Nachhaltigkeit (2013), applied on smart electric car

A "smart electric car" is particularly beneficial, if its use provides comfort by allowing entry into environmental protection zones, use of particular lanes etc., thus further enforcing the vision of facilitated business initially expressed within Fig. 1. Corresponding concepts can nowadays already be found abroad, e.g. in Norway (Cars 21 2013).

Besides facilitating the use of the electric car, holistic offers of ICT services may increase resale value of the car, particularly, if second-hand customers can modify and purchase IT services to their particular business requirements, conduct updates etc.

As a matter of fact, battery technology will still be a major driver as with regard to resale price development. Therefore, flexible pay-per-use offers for the battery alone will enable companies to overcome value-loss anxieties. Since the battery stays in possession of car producers/sellers, overhauling/refresh strategies e.g. by replacing single cells may prove economically feasible when a certain threshold of batteries in use is reached. Information necessary for the company to access the state of the battery and decide about second hand use or recycling can be withdrawn from the data collected during use and be used within ICT services developed.

To put it in a nutshell, individual assessment of the company's fleet requirements, conduction of modifications to the car both from engineering and ICT service point of view as well as service facilitating vehicle resale can provide added value to the customer and improve TCO calculations. To which extent this will happen, depends on the individual use modalities and can be part of the initial needs assessment of the company. Besides serving the customer, data can also be processed by the manufacturer to analyze both conditions of the electric car and ensure efficient service and resale.

## 5 Conclusion

As pointed out, companies put strong emphasis on cost and image considerations when purchasing fleet cars. These can be boiled down to five major criteria—adding "Ecology", "New products and services" and "Facilitation of business processes" as points of concern.

Although properties of the electric car allow for several particular applications and current market development may provide additional benefits, the five criteria identified here have not yet been fully addressed. Only a small number of electric cars, particularly battery electric cars, have been sold in Germany until the end of 2013. Since the variety of models available for purchase is slowly broadening as premium manufacturers enter the market, image issues are being tackled and might soon be solved. As with regard to cost, solutions are not yet visible.

Therefore, services tackling cost disadvantages from a total cost of ownership perspective need to be considered. Currently, offers mainly comprise consultancy service, such as fleet assessment, in order to determine application opportunities. Besides, new financial models, such as battery leasing, seek to address high upfront costs of purchase by spreading them over the period of use. By shifting the risk of early battery damage from the customer to the manufacturer or retailer, commercial risk is largely eliminated.

However, there is currently no holistic service approach tackling cost over the entire life-cycle of the electric car. The paper has pointed out a variety of options, from tailored manufacturing to services facilitating a company's electric fleet business during mid-of-life and end-of-life phase. Particularly the latter is considered highly challenging in implementation due to the substantial amount of data that needs to be collected and processed from a variety of resources.

To date, the smart car business model mainly focus on the private driver, his comfort and safety. However, first concepts shown at automotive exhibitions indicate that connectivity and inclusion of multi-data sources will play a key role in provision of new services also for company delivery service and company fleet vehicles.

By the end of 2013, these services did not take into account the purpose of the ride and therefore lack inclusion of corresponding parameters.

Potential for attractive services in this domain are huge, particularly in the context of application in mega cities, where traffic jams cause serious delays and decrease quality of service.

The economic effects of the measures suggested are hard to estimate. A practical assessment involving design and testing of prototype smart electric cars forms a next important step in research to verify preliminary assumptions as expressed in Fig. 2.

**Acknowledgements** The research findings expressed within the paper were elaborated in the context of the Research project "New Mobility in Rural Areas" (NeMoLand) funded by the German Federal Ministry of Transport and Digital Infrastructure from 2011 to 2014.

# References


BME (2000) Fuhrparkmanagement – Konzepte für Planung, Finanzierung und Fahrzeugeinsatz. Huss-Verlag, München

Cars 21; online information platform. http://www.cars21.com/news/view/5566. Accessed 25 Oct 2013

Corporate Vehicle Observatory (2012) Praxistauglichkeit und Zukunftsaussichten von Elektro- und Hybridfahrzeugen. http://www.arval.de/ger/full-service-leasing/unternehmen/news/pressemitteilungen/arval-elektromobilitaet-18092012.html

Fuhrpark+management (2012) SUV auf dem Vormarsch. http://www.fuhrpark.de/suv-auf-dem-vormarsch/150/1801/57011/

Holzer N (2012) in Firmenauto: Fuhrparkmanager – kein Job wie jeder andere. http://www.firmenauto.de/vom-flottenchef-zum-mobilitaetsmanager-berufsbild-fuhrparkmanager-4441272.html. Accessed 18 Sept 2013

In Hannoversche Allgemeine (2013) EU-Kompromiss: Anreize für Elektroautos. http://www.haz.de/Nachrichten/Wirtschaft/Deutschland-Welt/EU-Kompromiss-soll-Autobauer-zum-Klimaschutz-zwingen. Accessed 18 Sept 2013

Kampker A et al (2013) Elektromobilität Grundlagen einer Zukunftstechnologie, p 103 ff. Springer, Berlin. ISBN 978-3-642-31985-3

Kraftfahrt-Bundesamt (2013) Bestand – Bestandsbarometer am 1. Januar 2013. http://www.kba.de/cln_031/nn_125398/DE/Statistik/Fahrzeuge/Bestand/2013_b_bestandsbarometer_teil2_absolut.html. Accessed 18 Sept 2013

Krischun S (2010) Total cost of ownership: Bedeutung für das internationale Beschaffungsmanagement. Diplomica Verlag, Hamburg

Lexikon der Nachhaltigkeit. http://www.nachhaltigkeit.info/artikel/1_3_a_drei_saeulen_modell_1531.htm?sid=1e43e3c7f637cf4b352228cce514e9d7. Accessed 25 Oct 2013

manager magazin online (2013) http://www.manager-magazin.de/unternehmen/autoindustrie/a-886657.html. Accessed 18 Sept 2013

Meschtscherjakov A (2009) Acceptance of future persuasive In-Car interfaces towards a more economic driving behaviour. In: Conference proceedings automotive UI 2009, Sept 21–22, 2009, Essen, Germany

Nationale Plattform Elektromobilität (2011) Zweiter Bericht der Nationalen Plattform Elektromobilität, p 43. http://www.siemens.de/elektromobilitaet/pdf/NPE-Zweiter-Bericht.pdf. Accessed 18 Sept 2013

Opel Webpage. http://www.opel.de/fahrzeuge/modelle/personenwagen/adam/index.html. Accessed 28 Oct 2013

Rother FW, Eisert R (2013) in Wirtschaftswoche: Stromer und Hybride erobern die Fuhrparks. http://www.wiwo.de/unternehmen/auto/fuhrparkmanagement/flottenautos-steuerliche-entlastung-fuer-privatnutzung/8010506-4.html. Accessed 18 Sept 2013

Sun J et al (2009) Context aware smart car: from model to prototype. J Zhejiang Univ Sci A 1049–1059. doi:10.1631/jzus.A0820154

Telekom Mobility Solutions Webpage (2010) DeTeFleetServices setzt Green Car Policy um. http://www.telekom-mobilitysolutions.de/dtag/cms/content/DeTeFS/de/1616030. Accessed 18 Sept 2013

Umweltbundesamt (2013) Kurzfristig kaum Lärmminderung durch Elektroautos. http://www.umweltbundesamt.de/laermprobleme/publikationen/position_kurzfristig_kaum_laermminderung_im_verkehr.pdf. Accessed 19 Sept 2013

Vivaldini M (2012) Improving logistics services through the technology used in fleet management. J Inf Syst Technol Manage 9(3):541ff. doi:10.4301/S1807-17752012000300006

Wirtschaft pro Klima; http://www.wirtschaft-pro-klima.de/Mobilitt. Accessed 18 Sept 2013

## Author Biography

**Jasmin Nehls** Dipl.-Wi.-Ing. Jasmin Nehls worked as research scientist at the Institute for Integrated Product Development at the University of Bremen until 2014. She then joined TU Braunschweig. Her work focuses on business models for sustainable technologies.

# Flexible Carsharing—Potential for the Diffusion of Electric Mobility

**Sandra Wappelhorst, Julia Dobrzinski, Andreas Graff, Josephine Steiner and Daniel Hinkeldein**

**Abstract** Electric mobility is a key to a sustainable, climate and environmentally friendly mobility. However, in terms of electrically operated cars it has not spread to the full extent of its potential. From the user's perspective there are two major barriers in buying an electric car: it costs more than a car with an internal combustion engine and also, it has in general a limited range compared to a conventional vehicle. Carsharing with electric vehicles is one promising solution to balance the mentioned high costs and the lack of range: Firstly, the total costs of ownership can be spread among many users and secondly, as a part of the public transport system it can offer a complement for local and long-distance travel. To validate that especially flexible carsharing services with electric cars are suitable in spreading electric mobility, the paper presents a study which was carried out in the city of Berlin comprising several surveys. Expectations and experiences of users with a flexible carsharing offer including exclusively electric vehicles were evaluated and analysed over a period of 1 year. Overall, the results show that most users of flexible e-carsharing have no problem with the specific characteristics of the electric vehicles. Driving range and charging are generally not a limiting factor. For the majority of the users the "e"-component is even a reason to use the offer. Thus, flexible e-carsharing seems to be a good concept to spread electric mobility.

S. Wappelhorst (✉) · J. Dobrzinski · A. Graff · J. Steiner · D. Hinkeldein
InnoZ—Innovation Centre for Mobility and Societal Change, Torgauer Straße 12–15, 10829 Berlin, Germany
e-mail: sandra.wappelhorst@innoz.de

J. Dobrzinski
e-mail: julia.dobrzinski@innoz.de

A. Graff
e-mail: andreas.graff@innoz.de

J. Steiner
e-mail: josephine.steiner@innoz.de

D. Hinkeldein
e-mail: daniel.hinkeldein@innoz.de



D. Fornahl and M. Hülsmann (eds.), *Markets and Policy Measures in the Evolution of Electric Mobility*, Lecture Notes in Mobility,
DOI 10.1007/978-3-319-24229-3_5

**Keywords** Electric mobility · Carsharing · Flexible carsharing · Electric carsharing · e-Carsharing

## 1 Introduction

Transport volumes in cities and towns around the world continue to grow as a result of economic growth, population increase and urbanisation. Peak oil, global warming, local emissions, traffic jams or the lack of urban parking spaces are major challenges for the transport sector on a global, national and local level. In this respect, innovative and integrated mobility concepts are required in meeting the objectives of economic competitiveness, social cohesion and balanced sustainable development (OECD and ITF 2012).

Electric vehicles (EVs) are believed to be a major help tackling the problems mentioned above. However, vehicles with an internal combustion engine (ICE) are still perceived as the universal cars from the majority of people. Manufacturers consequently pursue their traditional concept of a "car": Thus, the majority of EVs are usually designed in the so-called "conversion design" implying that the propulsion of vehicles originally designed with an internal combustion engine is simply replaced by electric propulsion. In the rare case of "purpose design" the vehicles are designed from scratch with electric propulsion (Ried 2013). However, the traditional concept of the so-called "fast, long-range sedan"—with a higher driving range due to a larger "energy reserve" compared to an electric vehicle—is still firmly anchored in the minds of many people (Canzler and Knie 2011). The vehicle with an ICE still represents a relatively autonomous and individual form of transportation (Scherf et al. 2010). In addition, from the user´s perspective there are two major obstacles in buying an electric car: First of all, an EV costs more than a current car with an ICE and secondly, electric cars have in general a limited range of approximately 100–200 km compared to a conventional car with a range of about 800 km (each depending on the vehicle type).

Integrated electric Mobility Services (IeMS) including carsharing with electric cars are one promising solution to face these barriers: The total costs of ownership can be spread among many users instead of only one owner and as a part of the public transport system electric cars within a carsharing fleet can offer a complement for local and long-distance travel. This double integration can balance the high costs of ownership and the lack of range.

Carsharing in general has become an important element of transportation systems worldwide aiming at decreasing car traffic and stimulating sustainable car use. As the name suggests, carsharing allows the use of vehicles on demand. After joining a carsharing organisation, individuals or businesses can rent a car without having the costs or responsibilities of ownership (Schäfers 2013). Carsharing is most common in urban areas where transportation alternatives are easily accessible. Nowadays, there are two different sharing models: Station-based offers with fixed parking spaces, return of vehicle to the station where it was rented and booking in advance. The fixed parking spaces are often situated near public transport interconnections such as railway stations in order to guarantee access to targets beyond mass transit. In contrast to station-based offers more recent services provide so-called flexible offers with no fixed stations. The cars are distributed within a designated business area of a city offering spontaneous booking and instant access, pay as you go, one-way and open-end-trips (Canzler and Knie 2009, 2013; bcs 2013).

Carsharing on a private level has been practised long before the first commercial organisations were established. The first documented carsharing organisation started in Switzerland in 1948. In the 1970s, there were several attempts in Europe to establish carsharing systems but it took until the 1990s before it gained worldwide popularity. Today, carsharing companies operate in about 1,100 cities, 27 nations and on five continents (as of August 2012) (Shaheen and Cohen 2013).

Research suggests the positive effects of carsharing. One major impact is the reduction in car ownership as some members sell their car after joining a carsharing organisation, avoid or postpone a vehicle purchase. Also, some studies indicate that members drive less after joining a carsharing organisation, thereby reducing the vehicle-kilometres travelled. Both, reduced vehicle ownership and vehicle-kilometres travelled contribute to lower greenhouse gas emissions, less congestion, better air quality and more public spaces. Furthermore, carsharing has also beneficial effects in terms of social inclusion: Households with no car can gain or maintain vehicle access without bearing the full costs of car ownership (Steininger et al. 1996; Burkhardt and Millard-Ball 2006; Martin et al. 2010; Firnkorn and Müller 2011; bcs 2013; Shaheen and Cohen 2013).

While the positive outcomes of carsharing with conventional cars have been evaluated intensively, less attention has been paid to electric carsharing fleets, their special characteristics and their potential in spreading electric mobility. To address this research gap, this article investigates the characteristics and behaviour of carsharing members with regard to EVs in general and flexible carsharing offers in particular, targeting local, regional and national research bodies as well as decision makers of the transport sector, industry, politics and the municipalities concerned.

## 2 Carsharing in Germany and the City of Berlin

Carsharing in Germany has a long tradition. The first official carsharing organisation “StattAuto Berlin” started in the year 1988, followed by schemes in the cities of Aachen, Bremen and Freiburg. Since then, the carsharing market in Germany has benefited from continuously growing memberships and the introduction of new carsharing organisations and services offering flexible systems and sustainable vehicle types with electric engines. By January 2013, there were more than 453,000 people using station-based services in 343 cities and communities as well as flexible carsharing in seven cities throughout Germany. On an international comparison,

**Table 1** Selection of carsharing offers in Berlin

| | Flinkster | E-Flinkster | DriveNow | Car2go | Multicity Carsharing |
|---|---|---|---|---|---|
| Type | Station-based | Station-based | Flexible | Flexible | Flexible |
| Booking | In advance | In advance | Spontaneous | Spontaneous | Spontaneous |
| Return | After booking time has expired | After booking time has expired | Parking when no longer required | Parking when no longer required | Parking when no longer required |
| Propulsion type | Combustor | Electric | Combustor (since 09/2011) and electric (since 06/2013) | Combustor (since 04/2012) and electric (since 12/2012) | Electric |
| In Berlin since | 2001 | 2010 | 09/2011 | 04/2012 | 08/2012 |
| Vehicles in Berlin | About 350 | About 50 | About 500 (including 40 EVs) | About 1,200 (including 20 EVs) | About 350 |
| Operator | Deutsche Bahn | Deutsche Bahn | BMW and Sixt | Daimler and Europcar | Deutsche Bahn and Citroën |

*Source* Deutsche Bahn (2013), DriveNow (2013), car2go (2013), Multicity Carsharing (2013). As of August 2013

carsharing in Germany is both in absolute terms and in proportion to the total population in second place (bcs 2013).

The largest carsharing network is provided by the German railroad company Deutsche Bahn. Since 2001 Deutsche Bahn offers carsharing under the brand name Flinkster a station-based offer in more than 140 cities in Germany. Since 2010, electric cars supplement the carsharing fleet of Deutsche Bahn under the name e-Flinkster. Reacting on the emerging market of flexible carsharing, Deutsche Bahn introduced in cooperation with Citroën its first flexible carsharing scheme in the city of Berlin in 2012, with initially 100 on-street parking locations. This was the third offer in Berlin after DriveNow and car2go to add flexible carsharing. In contrast to the other two organisations Deutsche Bahn and Citroën started out with solely battery electric vehicles under the brand name Multicity Carsharing (Hinkeldein et al. 2012). Table 1 summarises the organisations and their specific service characteristics mentioned above.

## 2.1 The Project Berlin Elektromobil

The e-Flinkster and Mulitcity Carsharing fleets are part of the research project Berlin elektroMobil (short BeMobility) which focuses on the intelligent networking of electric shared vehicles and public transport, exemplified by the city of Berlin. One aim of the first phase within the project BeMobility 1.0 (duration 2009–2011) was the integration of the electric carsharing system e-Flinkster in the conventional station-based Flinkster fleet and its linkage with Berlin's public transport system. After the successful completion of the project, the follow-up project BeMobility 2.0 started in 2012 for another 2 years.

Based on the findings of BeMobility 1.0 the existing station-based carsharing fleet with the electric vehicles e-Flinkster was optimised and extended. One project module was the installation of the new electric flexible carsharing offer Multicity Carsharing in the city of Berlin. The accompanying research consisted of several surveys identifying the acceptance of the different offers as described above.

The long-term goal of the projects was the dissemination of multimodal and integrated mobility services for passenger travel on a larger geographical perspective (Bock et al. 2012; Hoffmann et al. 2012).

Altogether, about 25 partners were involved in the two projects, including the energy sector, solar power industry, automotive supply companies, information service providers and transport operators. The research was carried out by the Innovation Centre for Mobility and Societal Change (InnoZ). The projects were funded by the German Federal Ministry of Transport and Digital Infrastructure (BMVI).

The article presents and discusses results of three quantitative surveys within the accompanying research of the project BeMobility 2.0. Considering the expansion of EVs in carsharing fleets, the article focuses on the question whether flexible carsharing offers with EVs can contribute to the diffusion of electric mobility. This question is analysed in-depth giving the example of Multicity Carsharing. To validate that flexible carsharing services are suitable in spreading electric mobility, the flexible Multicity Carsharing is also compared with the station-based e-Flinkster.

## 2.2 *e-Flinkster and Multicity Carsharing in Berlin*

Like Flinkster, e-Flinkster is a station-based offer meaning that if a car is no longer required it has to be returned to the station where it was rented. Booking needs to be done in advance, the price depends on time (billed per hour), kilometres travelled and vehicle type. The stations are mainly situated at public transport connections within the Berlin S-Bahn circle which covers the city center of Berlin with an area of about 88 $km^2$ (Berlin 892 $km^2$) where about 1 million of the 3.5 million citizens of Berlin live (as of 2012). e-Flinkster offers three different vehicle classes including six electric vehicle types. As a member of Flinkster it is possible to use the conventional and electric cars as well as the cars of Multicity Carsharing.

In contrast to Flinkster, Multicity Carsharing provides instant access, open-end and one-way-trips. The cars can be parked within the defined business area which also covers the Berlin S-Bahn circle and adjacent areas. The fleet consists of the car model "Citroën C-Zero" a purpose-designed electric car.

When Multicity Carsharing was introduced, the business area was initially marked by the S-Bahn circle including the exhibition site. The journey time was accounted at 10 min intervals with a daily maximum rate of 39.00 € exclusive parking costs. Due to the results of the accompanying research the offer was adapted in April 2013: Since then the journeys are billed by the minute, the business area has been extended and the vehicle fleet has increased from initially 100–350 cars. In addition, the registration fee has been reduced and prepaid credits complement the tariff system with daily rates of a maximum of 39.00 € inclusive parking costs. Also, unlike before, starting a charging process is credited with free driving minutes. In addition, customers of Multicity Carsharing can use further mobility offers: For instance, they have instant access to the whole Flinkster network in Germany via the Multicity Carsharing account and the membership card.

## 3 Methods

To answer the question whether flexible carsharing offers with EVs like Multicity Carsharing can contribute to the diffusion of electric mobility, results of the accompanying research within the project BeMobility are presented and discussed.

The surveys relevant for discussion concentrate on three quantitative surveys of the project BeMobility 2.0. The first survey focused on the expectations towards Multicity Carsharing prior to the first use ($T_0$-survey) and was carried out over a period of 3 weeks starting in August 2012. Target group were Flinkster customers of Berlin and the surrounding area (within a circle of 10 km). The aim of the second survey ($T_1$) was to evaluate the first experiences with Multicity Carsharing after half a year on the market and was conducted from November 2012 until January 2013 targeting Flinkster and Multicity Carsharing customers in Berlin, who had used the offer at least once. To analyse the long-term experiences with Multicity Carsharing including the adaption of the service as described above a third survey ($T_2$) was carried out from July until September 2013. Here, the sample consisted of Flinkster and Multicity Carsharing customers in Berlin who had used the offer at least once or who had been taking part in the $T_0$-survey.

All three surveys consisted of several sections, including questions about socio-demographics, current travel patterns and carsharing-related issues such as expectations, experiences and use intentions with regard to Multicity Carsharing in general and the car model Citroën C-Zero in particular. The surveys were developed as a computer assisted web interview (CAWI) and distributed via mail. The addresses of the target group were provided by the customer data base Flinkster and Multicity Carsharing of Deutsche Bahn.

In the following chapter, the data of the three surveys are also compared to the final study of BeMobility 1.0 ($T_3$-survey) where necessary. The survey evaluated the long-term experiences with e-Flinkster after 2½ years of its introduction (longitudinal analysis). The study was carried out over a period of 1 month starting in June 2012.

For the statistical evaluation, the ratings of the attitudes and the quantitative surveys were mainly closed questions using a six-point Likert scale. These were used depending on the context for approval or rejection (1—"not correct at all"; 6—"fully correct"). Also, established scales of the representative study "Mobilität in Deutschland" (MiD) were used to evaluate the mobility of the respondents ["(almost) daily"; "1–3 days/week"; "1–3 days/month"; "less than monthly"; "(almost) never"]. In addition, qualitative parts were used to validate and specify the quantative results and to provide reasons for the different assessments.

The statistical analyses generally refer to the calculation of the arithmetical average or the standard deviation. Also, multivariate statistical methods were used such as cluster analyses for the development of attitude-based mobility typologies. For the statistical correlation the correlation coefficient of Pearson's r was used. The corresponding significance tests were carried out on the level of significance 5 %.

# 4 Results

The following analyses focus on the participants' socio-demographics, their travel behaviour, their assessment of different carsharing offers and their experiences with electric carsharing.

In total, 379 people from Berlin responded to the $T_0$-survey, 150 to the $T_1$-survey and 492 persons answered the $T_2$-questionnaire. In addition, 79 people took part in the $T_3$-survey.

During the different stages of the survey implementations the carsharing market in Berlin developed very dynamically: At the time of the $T_0$-survey the station-based Flinkster was available offering conventional and electric cars. In addition, DriveNow and car2go had already introduced their flexible carsharing with merely conventional cars at that time. The flexible Multicity Carsharing with initially 100 EVs (nowadays 350 cars) was just about to start. During the $T_1$-survey car2go added 20 EVs to their fleet. When the $T_2$-survey was conducted DriveNow also introduced electric cars with 40 by number (Table 1).

## *4.1 Carsharing Users and Their Travel Behaviour*

All three surveys show a high rate of male respondents (ranging between 84 and 91 %) with a higher education level (82–86 %). The majority of the sample size works full time (76–79 %) and has a monthly household net-income over 4,000 Euro (38–40 %). Also, the majority lives within the business area of Multicity Carsharing (71–85 %).

In terms of mobility type, the respondents are mostly innovative technic-orientated multioptional persons, having a high awareness regarding the different transport modes, positive attitudes towards environmental issues and mobility offers as well as a high rate on the innovator scale. Another high proportion of the sample are so-called ecologically sensitive persons with a high orientation towards public transport and cycling. They prefer public transport modes and cycling and have no particular passion in driving a car. Also, they have a positive attitude towards environmental issues and a positive rate on the innovator scale, preferably living in an urban environment. Within the $T_2$-survey, 51 % of the respondents belong to the first group of innovative technic-orientated multioptionals, 37 % to the second group. Both groups are in general predestinated for new and integrated mobility solutions like different forms of carsharing including conventional and EVs.

In terms of travel behaviour, the results of the surveys indicate that the introduction of EVs in flexible carsharing fleets results in an increased use frequency. While carsharing (as driver and co-driver) was initially used by 14 % of the customers between once a week and daily ($T_0$, n = 372), the proportion has almost doubled to 26 % after flexible carsharing has been introduced to the carsharing market in Berlin ($T_2$, n = 449). The same developments can be seen for carsharing with EVs: Prior to the introduction of Multicity Carsharing only 1 % used e-carsharing once to three times a week ($T_0$, n = 370). This proportion has increased to 14 % using e-carsharing once to three times a week up to a daily basis. Even though there were EVs in place with the station-based e-Flinkster in Berlin, the flexible forms of Multicity Carsharing, car2go and DriveNow seem to have increased its use significantly.

The comparison of journey time and trip purpose reveals that Multicity Carsharing is usually used for spontaneous trips and every day travel purposes. In contrast, the station-based e-Flinkster is generally booked for planned, longer trips (in terms of duration and distance) and leisure purposes. This arguably indicates that use patterns adapt to the boundary conditions of certain mobility services. Also, the results give first indications that flexible carsharing offers might have the potential to spread electric mobility.

## 4.2 *Business Models of Carsharing*

As shown above, the carsharing market in Germany and Berlin has developed very dynamically within the last few years including a variety of carsharing providers offering different service characteristics. To collect the opinions with regard to different business models (Table 2) the participants were given four semantic differentials: Innovative and conservative, practical and impractical, simple and complicated, stylish and not stylish.

The results indicate that the flexible carsharing systems are clearly estimated as innovative whereas the station-based schemes are seen as conservative, regardless the propulsion type (Fig. 1). In terms of practicability and simplicity/complexity the users rate the flexible systems more practical than the station-based systems.

**Table 2** Selected business models of the carsharing sector

| | A | B | C | D |
|---|---|---|---|---|
| | Station-based carsharing including cars with ICE | Station-based carsharing including EVs | Flexible carsharing including cars with ICE | Flexible carsharing including EVs |
| Rental/return point | Station-based | Station-based | Flexible | Flexible |
| Booking | In advance | In advance | Spontaneous | Spontaneous |
| Return | After booking time has Expired | After booking time has expired | Parking when no longer required | Parking when no longer required |
| Propulsion type | Combustor | Electric | Combustor | Electric |

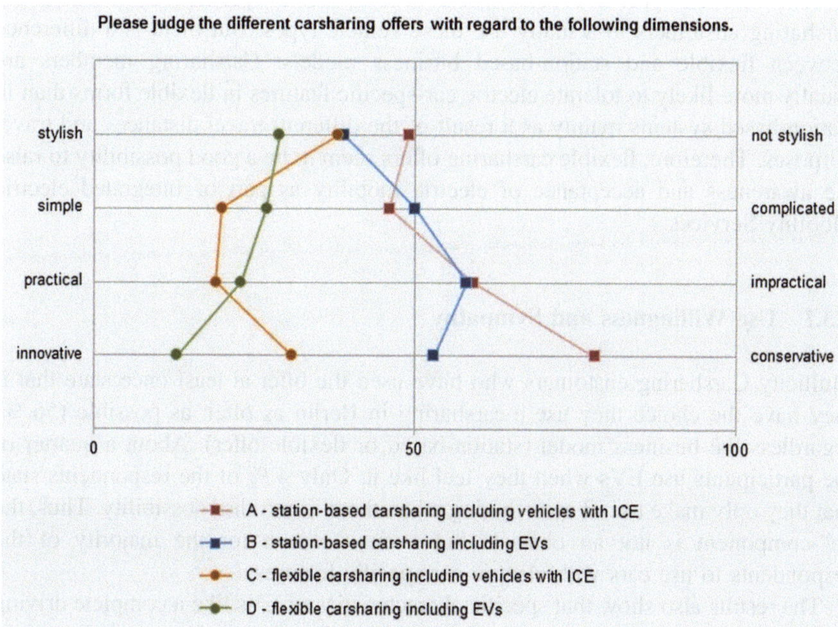


**Fig. 1** Evaluation of different Carsharing offers

Obviously, flexible forms of carsharing combining spontaneous use, no restrictions regarding time and the possibility to return the vehicle when no longer required at a dedicated on-street parking location are practical and can be integrated easier in the daily mobility routines than station-based offers. Also, users believe that the flexible carsharing with electric cars is almost as practical and easy as the offers with a combustor. This is a first indication that use barriers of e-carsharing in terms of "e"-components might not apply for certain customer groups.

## 4.3 *Flexible e-Carsharing*

In general, use barriers of electric vehicles in private ownership include the high costs of purchase, a perceived limitation in mobility, uncertainty about the battery as an energy storage (Müller et al. 2011), the short driving range, the lack of a sufficient charging infrastructure, long charging times of the battery (Proff 2013) and the uncertainty about the future resale value of an EV (Agassi 2009). In terms of EVs as part of carsharing fleets it is usually assumed that factors like the limited driving range and the complicated charging process are the main barriers for customers in using such offers.

However, the results of the different surveys within the project BeMobility indicate that the specific characteristics of EVs are not particularly a barrier for carsharing customers to actually use these vehicle types. But there is a difference between flexible and station-based business models: Carsharing members are usually more likely to tolerate electric car-specific features in flexible forms than in station-based systems mainly as a result of the different travel distances and travel purposes. Therefore, flexible carsharing offers seem to be a good possibility to raise the awareness and acceptance of electric mobility as part of Integrated electric Mobility Services.

### 4.3.1 Use Willingness and Sympathy

Multicity Carsharing customers who have used the offer at least once state that if they have the choice they use e-carsharing in Berlin as often as possible (56 %) regardless the business model (station-based or flexible offer). About a quarter of the participants use EVs when they feel like it. Only 4 % of the respondents state that they only make use of e-carsharing when there is no other possibility. Thus, the "e"-component is not an obstacle but even a reason for the majority of the respondents to use cars with electric car-specific features.

The results also show that specific characteristics of EVs like a complete driving range and a 100 % battery level are rather secondary for the interviewed users of Multicity Carsharing. The most frequent reasons for not using Multicity Carsharing are the availability of vehicles and the small size of the operating area. For the majority of the respondents the one-way characteristic and the "e"-component of Multicity Carsharing is even a reason to use this offer (Fig. 2).

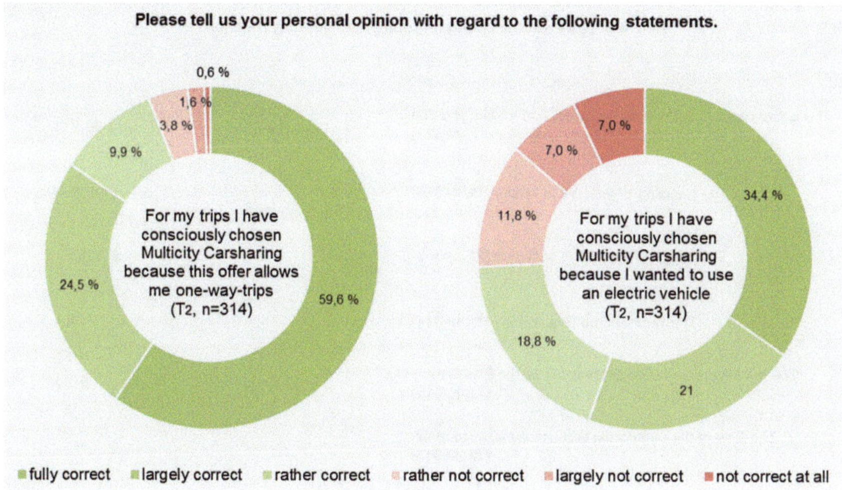


**Fig. 2** Reasons for choosing Multicity Carsharing

### 4.3.2 Suitability for Everyday Use

The handling of EVs should be easy for customers in order to meet and satisfy their daily mobility routines. This aspect is essential for the acceptance of electric mobility offers in general and electric cars in carsharing fleets in particular.

The evaluation of the everyday suitability with regard to Multicity Carsharing shows that the specific characteristics of EVs do not cause any problems when using Multicity Carsharing (Fig. 3). All system components like battery level, handling and range of the electric cars, the return process, the booking/reservation and the paying system are judged on a scale from 1 ("not correct at all") to 6 ("fully correct") with values between 4.78 and 5.36.

In general, the respondents agree that the electric car-specific components of vehicles in carsharing fleets meet their daily mobility routines which have in turn positive implications for the use acceptance of such offers.

### 4.3.3 Insecurities

Even though the everyday suitability is judged positive, insecurities might contradict these viewpoints. However, the majority of the customers feel safe regarding reliability of the car, finding a charging point, parking within the business area, fluctuation of the range indicator, the charging state and level of the battery and range (Fig. 4). All aspects are evaluated on a scale from 1 ("not correct at all") to 6 ("fully correct") with positive average values between 1.17 and 2.33.

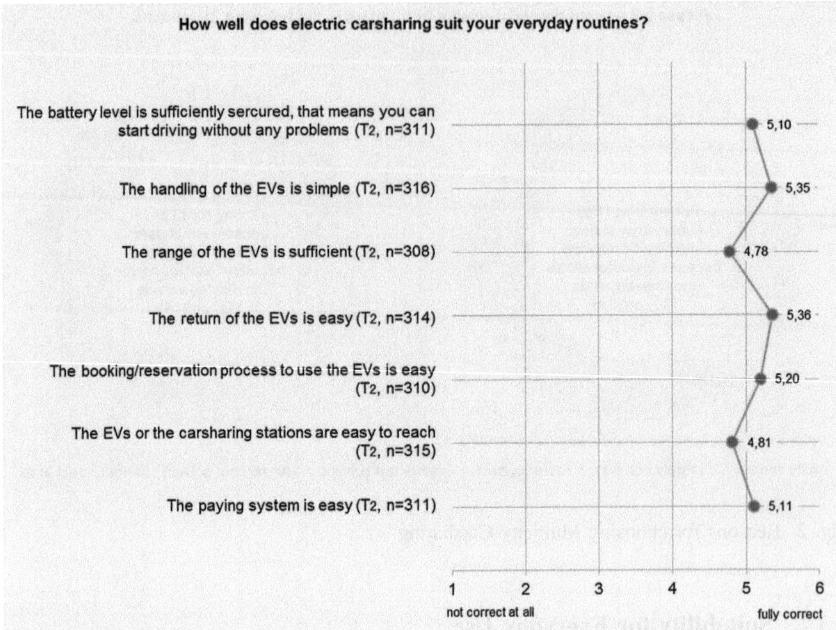


**Fig. 3** Everyday suitability of EVs (mean values)

The main factors for discomfort are unforeseen longer travel times (i.e. due to high volumes of traffic, looking for a parking space) and the corresponding additional costs which are rated with average values of 3.52 and 3.55.

Having a deeper look at the aspect of charging which is assumed to be one of the main barriers in using EVs in carsharing fleets the results indicate that charging plays a rather rudimentary role in the context of flexible e-carsharing: Only about one-third of the respondents ended their trips with a Multicity Carsharing car at a charging point, about two-third left their vehicle within the designated business area. The main reason leaving the car at a charging point was the availability of a free parking space. As trips are generally short, the necessity of recharging the car is comparably low. Also, cars can be parked anywhere within the designated business area without the necessity to end it at a charging point. The user is only obliged to ensure a vehicle driving range of at least 10 km when ending the trip. Since the adaption of the system in April 2013, users can receive free driving minutes when ending a trip at a charging point which has resulted in a higher percentage of people actually ending the trip at a charging point.

A frequently expressed criticism of the station-based e-Flinkster was the aspect of battery level. However, for the majority of people using the flexible Multicity Carsharing this is not a point of concern: 94 % of the respondents would be willing to accept a battery level of up to 50, 60 % of the participants would even accept a battery level of up to 30 % ($T_2$, n = 317).

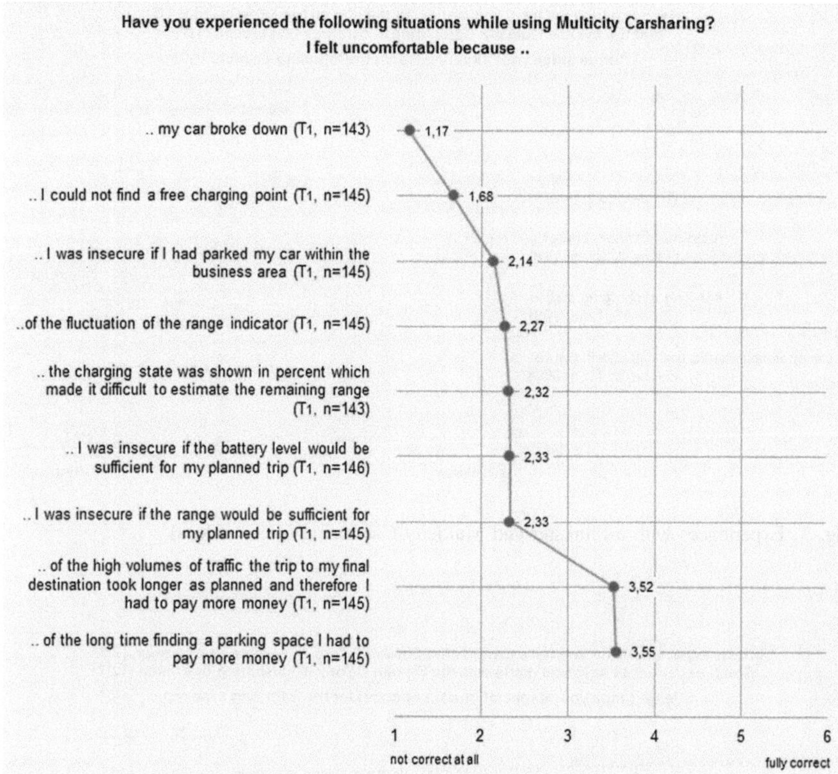


**Fig. 4** Insecurities with regard to flexible e-carsharing (mean values)

In terms of driving range, which is also believed to cause insecurities the results indicate that this aspect plays a much bigger role when using station-based offers like e-Flinkster than for flexible business models like Multicity Carsharing (Fig. 5).

These results do not surprise as station-based offers are usually used for longer trips whereas flexible carsharing systems are preferred for shorter distances which requires in turn less battery level.

### 4.3.4 Vehicle-Specific Characteristics

For the customer acceptance and satisfaction it is also important to analyse how the customers get along with the vehicle itself. As for Multicity Carsharing, the fleet only consists of the purpose-designed Citroën C-Zero, a four-seated car.

In general, the respondents fully agree that the use of the Citroën C-Zero is easy. This covers all distances travelled. Expectations and experience regarding the “e”-specific characteristics of the Citroën C-Zero show that in general the experiences with regard to certain aspects are much better than the expectations (Fig. 6) even though for both survey periods the results show positive values.

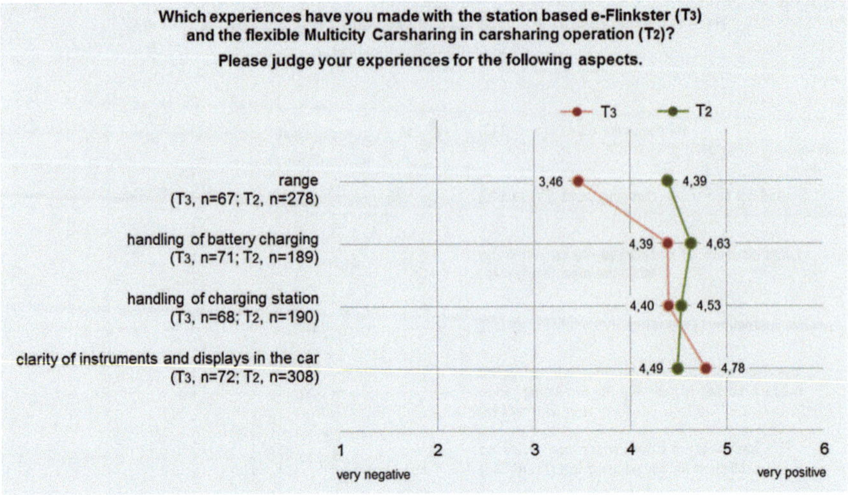


**Fig. 5** Experiences with e-Flinkster and Multicity Carsharing (mean values)

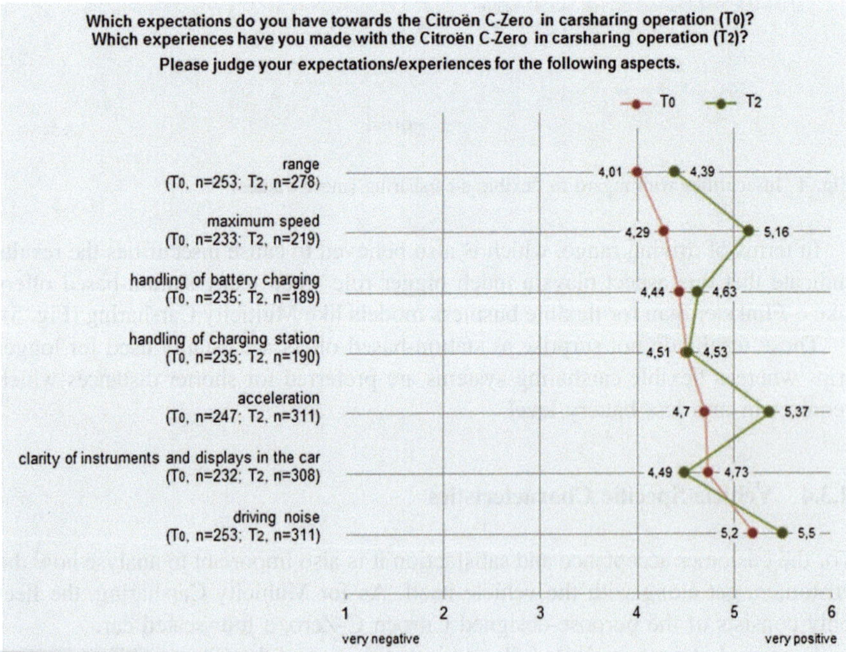


**Fig. 6** Expectations and experiences with the Citroën C-Zero (mean values)

Only the aspect of clarity of instruments and displays shows a slight decline after using the cars. Especially the maximum speed and acceleration exceed the expectations. But also the range, the handling of battery charging and the driving noise are evaluated better after the use. These results underline that direct and personal experience can minimise existing prejudices as far as they exist. Positive experiences can also contribute to a higher satisfaction and acceptance of electric cars in general and in carsharing fleets in particular.

### 4.3.5 Future Purchase and Use Intentions

In order to figure out if it is possible to integrate EVs in everyday mobility routines, the participants were asked if they are intending to buy a car in the near future. The majority of the respondents (96 %) answer in a negative way: For them, buying an electric car is not an option. Relatively low is also the willingness to replace an existing vehicle in favour of public transport or carsharing: 79 % of the respondents give a negative answer.

However, the majority of the participants state that in the future they will use an EV in rental systems. Here, the proportion of persons intending to use Multicity Carsharing is significantly higher than of people intending to use e-Flinkster.

In general, the interviewees take a positive view of the future potential and the long-term diffusion of electric mobility. They see its potential mainly in integrated concepts combining different transport modes like public transport or cycling. Moreover, people do not think that electric mobility is merely a temporary fashion but has future potential especially in combination with carsharing: The majority of the respondents think that EVs in carsharing fleets make more sense than in private ownership. In addition, the interest in EVs is high: 91 % state that they will gather information about electric mobility in the future.

Altogether, the participants have an optimistic view for the diffusion of electric mobility, especially in integrated mobility concepts and as part of carsharing fleets. In contrast, the opinions concerning electric cars in private use are clearly reserved.

## 5 Conclusion

Overall, the results of the different surveys within the research project BeMobility indicate that most users of the flexible Multicity Carsharing have no problem with the specific characteristics of the EVs. Especially the limited range is an often stated problem as for example the results of BeMobility 1.0 with station-based e-Flinkster or studies on private e-car users show. However, the hypothesis of range anxiety could be falsified for flexible e-carsharing.

For the majority of the respondents the "e"-component is even a reason to use the flexible Multicity Carsharing which also meets the daily mobility routines. Thus, flexible carsharing with EVs seems to be—in contrast to station-based offers—a good concept to spread electric mobility and to overcome the still existing barriers and prejudices when it comes to electric vehicles in private use.

Even though the perceptions, experiences and future use intensions are high when it comes to e-carsharing the willingness to actually buy an electric vehicle is low. Electric cars still compete with conventional cars and have major disadvantages especially when it comes to long distance travel. The range limitation due to the limited storage capacity of electric batteries is a major drawback for the diffusion of electric mobility. Furthermore, the costs of buying an electric car in private use are still much higher than of conventional cars.

Against this background, strategies for the successful diffusion of electric mobility should focus on the following aspects:

- Electric cars as part of new, intermodal and integrated transport networks have the potential to overcome barriers like the often discussed limited range. Especially flexible carsharing systems have specific advantages as distances travelled are generally shorter than in station-based offers. Here, the limited driving range and storage capacity can be neglected. Also, sharing models in general have the advantage that the higher costs of electric mobility in private ownership can be spread among many people.
- EVs need to be defined separately and designed differently to conventional cars in order to overcome the traditional concept of the so-called "fast, long-range sedan"—with a higher driving range due to a larger "energy reserve" compared to an EV. Purpose-designed cars like the Citroën C-Zero are steps in the right direction. The differentiation of traditional "cars" and new mobility forms including EVs is an important factor to spread electric mobility and to increase its acceptance in public.
- Electric mobility is often perceived as an ecologically sustainable means of transport. Therefore, the environment-friendliness of electric mobility needs to be secured by generating electricity from renewable energy sources.

To meet these challenges, the participation and cooperation of the public and private actors are crucial, including the mobility sector, industry, politics and science as well as the municipalities concerned.

**Acknowledgments** This study was kindly funded by the German Federal Ministry of Transport and Digital Infrastructure (BMVI) whose support made this research possible.

## Bibliography

Agassi S (2009) World without oil. Better place builds a future for electric vehicles. In: Innovations, fall 2009, pp 125–140

bcs Bundesverband CarSharing e.V (2013) Bundesverband carsharing annual balance 2012: carsharing growth as never before. http://www.carsharing.de/index.php?option=com_content&task=view&id=373&Itemid=44. Accessed 30 June 2013

Bock B, Wolter F, Scherf C (2012) BeMobility: integrated ecarsharing in Berlin. Routes-Roads 354:76–83

Burkhardt J, Millard-Ball A (2006) Who is attracted to carsharing? Transp Res Rec: J Transp Res Board 1986(1):98–105

Canzler W, Hunsicker F, Karl A, König U, Lange G, Maertins C, Ruhrort L (2007) DB Mobility—Beschreibung und Positionierung eines multimodalen Verkehrsdienstleisters, InnoZ Baustein Nr. 1, Innovationszentrum für Mobilität und Gesellschaftlichen Wandel, Berlin

Canzler W, Knie A (2009) Green solutions to the auto crisis—from auto makers to mobility service providers, strategy paper, Heinrich Böll Foundation (Editor), Publication Series on Ecology, Volume 4, Berlin

Canzler W, Knie A (2011) Einfach aufladen. Mit Elektromobilität in eine neue Zukunft, München

Canzler W, Knie A (2013) Schlaue Netze: Wie die Energie- und Verkehrswende gelingt, München

Car2go (2013) Mehr Berlin mit mehr als 1200 car2go. https://www.car2go.com/de/berlin/. Accessed 27 Aug 2013

Deutsche Bahn (2013) Flinkster - Mein Carsharing. http://www.flinkster.de/. Accessed 27 Aug 2013

DriveNow (2013) BMW und MINI in ganz Berlin. https://de.drive-now.com/php/metropolis/city_berlin?cit=6099&language=de_DE. Accessed 27 Aug 2013

Firnkorn J, Müller M (2011) What will be the environmental effects of new free-floating car-sharing systems? The case of car2go in Ulm. Ecol Econ 70(8):1519–1528

Hasse S, Hendzlik M, Hoffmann C, Hinkeldein D (2011) Customer needs and acceptance of electric vehicles as part of integrated transport concepts. In: Presented at the final conference of the socio-economic research of the "Model region electric mobility Bremen/Oldenburg", Bremen

Hinkeldein D, Hoffmann C, Schönduwe R (2012) Using attitude-based focus groups to analyze the potential of electric vehicles as part of integrated mobility services. In: Presented at the transportation research board, Washington

Hoffmann C, Graff A, Kuttler A, Hendzlik M, Wolter F (2012) Bewertung integrierter Mobilitätsdienste mit Elektrofahrzeugen aus Nutzerperspektive, InnoZ Baustein Nr. 11, Innovationszentrum für Mobilität und gesellschaftlichen Wandel, Berlin

Hoffmann C, Hinkeldein D, Graff A, Kramer S (n.d.) What do users think about electric mobility? In: Evolutionary paths towards the mobility patterns of the future. Springer, Berlin

Martin E, Shaheen S, Lidicker J (2010) The impact of carsharing on household vehicle holdings: results from a North American shared-use vehicle survey. Transp Res Rec 2143:150–158

Multicity Carsharing (2013) Citroën multicity carsharing Berlin. https://www.multicity-carsharing.de/. Accessed 27 Aug 2013

Müller C, Benad H, Rennhak C (2011) E-Moblity - Treiber, Implikationen für die beteiligten Branchen und mögliche Geschäftsmodelle. Reutlinger Diskussionsbeiträge zu marketing and management, no 09

OECD Organisation for Economic Co-operation and Development, ITF International Transport Forum: Transport Outlook (2012) Seamless transport for greener growth, Paris

Proff H (2013) Geschäftsmodelle zwischen technischen Herausforderungen und betriebswirtschaftlichen Notwendigkeiten im Übergang in die Elektromobilität. In: Proff H (ed) Herausforderungen für das automotive engineering and management. Springer Fachmedien, Wiesbaden, pp 1–23

Ried M (2013) Technik- und Kostengesichtspunkte in der Entwicklung elektrifizierter Fahrzeuge. In: Proff H (ed) Herausforderungen für das automotive engineering and management. Springer Fachmedien, Wiesbaden, pp 25–36

Schäfers T (2013) Exploring carsharing usage motives: a hierarchical means-end chain analysis. Transp Res Part A 47:69–77

Scherf C (2010) Entwicklungshürden der Elektromobilität. Das Verhältnis zwischen Antriebsparadigma und Automobilleitbild. InnoZ-Baustein Nr 8, Innovationszentrum für Mobilität und gesellschaftlichen Wandel, Berlin

Shaheen SA, Cohen AP (2013) Carsharing and personal vehicle services: worldwide market developments and emerging trends. Int J Sustain Transp 7(1):5–34

Steininger K, Vogl C, Zettl R (1996) Car-Sharing organizations: the size of the market segment and revealed change in mobility behavior. Transp Policy 3(4):177–185

Wappelhorst S, Graff A, Steiner J, Hinkeldein D (2013) New carsharing offers and customer groups: implications for a growing and diversifying market. In: Proceedings of the European transport conference (ETC) 2013, Frankfurt

## Author Biographies

**Sandra Wappelhorst** works as a Senior Expert at the Innovation Centre for Mobility and Societal Change in Berlin. She manages research projects dealing with the implementation of intermodal electric mobility services in urban and rural areas.

**Julia Dobrzinski** is a Student Research Assistant at the Innovation Centre for Mobility and Societal Change in Berlin. She earned her Bachelor's degree in General Business Administration at the Free University of Berlin and is currently studying as a Master student in Finance and Management at the University of Münster. Her research interests are Strategic Management, Innovation in the mobility sector and Sustainability Reporting.

**Andreas Graff** graduated from the Technical University Berlin (TUB) with a degree in Sociology and Technology Studies. He has worked as a research assistant at the chair of sociology at TUB where he taught methodology of scientific research and was involved in several projects concerning this topic. His work at the Innovation Centre for Mobility and Societal Change in Berlin focuses on method development and standardization for mobility research and consumers acceptance.

**Josephine Steiner** holds a degree in Transport Economics (Dresden University of Technology). Her research at the Innovation Centre for Mobility and Societal Change in Berlin focuses on sustainable mobility from a social scientific perspective. Josephine Steiner has been involved in projects dealing with consumer acceptance of transport services that integrate public transport and electric vehicles.

**Daniel Hinkeldein** works as a Project Manager at the Innovation Centre for Mobility and Societal Change in Berlin. He manages national and international industrial projects for energy suppliers and mobility providers to develop user-friendly electric mobility services.

# Part II
# Policy Perspectives

# Public E-Mobility Funding and its Influence on Cooperation and Innovation

**Isa Reinecke**

**Abstract** Large sums of public funding are provided to promote research, development and demonstration projects related to electric mobility. Especially collaborative projects are funded due to the assumption that cooperation enhances innovation. However, there is the risk of government failure. Therefore, public measures have to be evaluated and, if necessary, be adjusted. In a qualitative survey with different organizations involved in the innovation system of e-mobility, impediments as well as proposals for improvement were identified. The study focused on cooperative behavior, demonstrating that political measures have a great impact on it.

**Keywords** Public funding · Innovation policy · Government failure · Cooperation

## 1 Introduction

As electric mobility is seen as one of the most important future technologies in upcoming years, large sums of public funding are provided to promote e-mobility and innovations in this field (Bundesregierung 2011; JRC 2013). In Germany, a considerable amount of respective research and development is funded by national and European means usually financing collaborative projects. It is assumed that innovation is created through interactive learning processes, thus cooperation between actors with different backgrounds is required. However, government intervention is subject to discussion. There is, for example, criticism that the funding budget in Germany is too small and funding is provided too widely-spread (Dudenhöffer 2010). This paper aims to examine to what extent policy instruments affect collaborative projects and in what way they have an influence on the creation

I. Reinecke (✉)
Centre for Regional and Innovation Economics, University of Bremen,
Wilhelm-Herbst-Str. 12, 28359 Bremen, Germany
e-mail: isa.reinecke@uni-bremen.de



D. Fornahl and M. Hülsmann (eds.), *Markets and Policy Measures in the Evolution of Electric Mobility*, Lecture Notes in Mobility,
DOI 10.1007/978-3-319-24229-3_6

of innovations in general. It is based on a survey carried out with 26 representatives from organisations that have been involved in publicly funded collaborative projects concerning electric mobility. Above all, most interviewees estimate political actions positively. Nevertheless, there seem to be unexpected and unwanted effects, respectively, that result from funding conditions and the way in which financial means are distributed. Besides, other public interventions such as legal prohibition or creation of user advantages have an impact on innovation processes, as well. But under what circumstances can such measures be fruitful and enhance innovation? What can policymakers do in order to improve framework conditions for collaborations?

This paper is structured as follows: First, a brief overview of the theoretical background regarding the reasons for innovation policies and their impact on the generation of innovations is given. Then, the methodology applied in the empirical study is explained. The third section presents the results of the interviews illustrating the effects of different political measures from the viewpoints of the agents affected. Findings are finally summarized and evaluated in the conclusion. The latter also comprises recommendations for action in order to improve innovative strength.

## 2 Theoretical Background

Market failure is the common rationale for government intervention defined by economists, i.e. government interventions are required, if market outcomes do not lead to results which are appreciated regarding welfare (Montmartin and Massard 2013; Maier et al. 2006; Tokila et al. 2008). Thus, government intervention can be necessary to correct non-optimal allocations of resources. Reasons for market failure are diverse. Knowledge shows characteristics of public goods, namely non-rivalry and only partial excludability. That's why innovators cannot exclusively profit from their achievements. Profit-maximizing enterprises therefore underinvest in research and development (R&D). Other unwanted effects are externalities like knowledge spillovers and the existence of a monopoly. Furthermore, public intervention in terms of incentives is regarded to be justified when projects are too risky for single organizations, but their implementation is desired by society. However, there is also the opposite effect of duplication of R&D activities. Due to ever shorter innovation cycles and the emergence of innovation races, different organizations may work on the same problems and develop the same solutions which is inefficient. Another manifestation of market failure is the spatially uneven distribution of economic activities and innovation outcome caused by externalities and imperfect mobility of resources (Maier et al. 2006; Montmartin and Massard 2013; Metcalfe 1995). According to the mentioned assumptions, opportunities of realizing public innovation goals are manifold. This paper concentrates on the so-called hard policy instruments which directly affect the behavior of the respective recipients (Welsch 2005). These include, above all, financial means in terms of funding programs for R&D and demonstration projects as well as the protection of intellectual property and the provision of knowledge by financing fundamental research, respectively.

Metcalfe (2007) considers the traditional rationale for innovation policy to be flawed and brings forward the approach of systems failure (or systemic failure, OECD 1997). The latter is expressed in a lack of transformation of scientific knowledge into products as well as a shortage of diffusion of new technologies and procedures. According to this approach, market failure is an inadequate concept in knowledge societies since innovation, generated by firms, would preclude equilibrium and perfect competition. Externalities, from this point of view, are not per se negative. For example, information spillovers do not necessarily lead to underinvestment because organizations could even profit from those. Putting ideas in the public domain can have positive benefits in terms of stimulating differential knowledge creation. This refers to the idea of feedback mechanisms in the innovation process (Kline and Rosenberg 1986). Thus, Metcalfe even asks for an encouragement of knowledge spillovers and holds the view that patents should not be deployed too broadly because this would limit the ability of other organizations of being creative. The public good character is not seen as a big problem since a lot of background knowledge is required in order to being able to use new knowledge. So there is self-exclusion due to lacking skills and abilities. Knowledge flows should be managed by involved organizations which is already done in practice. Moreover, there are no futures markets for investments regarding innovation. As market prices do not convey relevant information, networks play an important role in mediating innovation processes. Collaborative R&D projects are another way of dealing with uncertain information. These forms of joint action are contrary to the idea of complete competition between isolated firms. The approach of systems failure shares a lot of common characteristics with the evolutionary approach of government intervention (Moreau 2004) which regards increasing the probability of experimental behavior to be the central concern of evolutionary politics.

Policy should aim at facilitating the emergence and self-organization of innovation systems. Therefore, missing components, connections and misplaced boundaries have to be addressed. This firstly means enhancing education of the population in order to make sure there are enough qualified employees. Furthermore, bridging arrangements between different organizations, e.g. universities and firms, have to be supported. As enterprises have to be able to cooperate fruitfully with universities to transfer knowledge into innovations, it can also be necessary to promote private research and development activities by tax credits, for example (OECD 1997; Metcalfe 2007).

Probably unconsciously, aspects of this approach are already being applied. Cooperation is one central component of current innovation strategies. Public measures often aim at research and development activities assuming that there is a linkage between innovation and economic growth (BMBF 2010). Following this logic, in the field of e-mobility policies try to support the creation of knowledge and innovations in order to promote technological progress, thereby enhancing competitiveness and generating economic growth. Considering innovations as products of cumulative learning processes, as it is also described in the chain-linked model of innovation (Kline and Rosenberg 1986), the generation of know-how is dependent on communication and interaction of different agents. Besides collaborative R&D projects there are incentives to encourage agglomerations of knowledge generating organizations in terms of science parks or cluster initiatives and to set up technology transfer offices. All those measures are what Metcalfe calls bridging arrangements. Montmartin and Massard (2013) also regard cooperation to be an important tool for minimizing the risk for either under- or overinvestment. Politics should therefore adopt measures to support investment in collaborative projects.

Nevertheless, there is also the risk of government failure. Following the regulations of the European Union (Europäische Kommission 2013), public intervention must not distort competition by, for example, favoring specific enterprises or branches or reducing private investment incentives. Moreover, there are personal interests such as being re-elected (Fritsch 2011). Bureaucracy is regarded as another source of government failure. Thus, bureaucrats would not be interested in realizing the best outcome, but in maximizing their budget. That's why all money provided is spent and every year more money is demanded. Since bureaucrats are not subject to market constraints there is no need to minimize costs (Fritsch 2011; Mitchell and Simmons 1994). Imperfect information, pressure group activity as well as political myopia are further sources of government failure (Moreau 2004). Public funding of (research) activities often implies the risk of deadweight, too. The latter means the "degree to which the project would have been carried out without grant assistance" (Tokila et al. 2008, p. 2). In case a project would have been carried out either way, public funding, being brought up by the tax payers, is wasted. Tokila et al. (2008) evaluated investment subsidies in Finland administrated by the Ministry of Trade and Industry. The authors found that deadweight is significantly smaller the higher the project costs are. It is also smaller for new firms than for old ones. Moreover, the likelihood of deadweight is lower in peripheral regions with higher unemployment and weaker economic growth rates. This may be due to higher difficulties of obtaining financing or loans. There have been several initiatives trying to support regions with poor economic performance. One example of regionally bound research and technology policy has been the subsidies for East German enterprises after German reunification. Deadweight effects seem to have been of no significance in this specific case, as firms hardly would have been able to carry out their innovation activities without funding (Czarnitzki 2001).

From a literature-based analysis, Montmartin and Massard (2013) draw the conclusion that insufficient private investment in R&D is not as numerous as expected. To their mind, public interventions should not just increase incentives for private investment, but rather concentrate on "the most efficient allocation of funds across industries, regions and types of research" (Montmartin and Massard 2013, p. 33). According to their study, surplus appropriability is the main explanation for under-investment. Therefore, they ask for a more demand-orientated policy orientation that encourages the purchase of innovations.

Taking into consideration that every product undergoes several phases before its establishment at the market, suitable instruments have to be chosen corresponding to the respective period. According to Fornahl and Diller (2010), there are three relevant stages: kick-off, growth and market penetration. Based on then available projections, they expected the period of growth in electric mobility to begin in 2015. Most measures have to be taken during the stage of kick-off and the beginning of the growth phase. First, the necessary infrastructure and the regulatory framework have to be built up. The first stage could also imply the encouragement of production by public participation in new business models like car-sharing. Public research funding is an important tool to promote e-mobility right from the beginning. Another incentive could be to grant users of e-vehicles certain advantages like free parking space, in particular during the first stage. To gain credence, e.g. by incorporating renewable energies, is an important task during the first two stages of market introduction. Qualification of professionals at universities and vocational schools is needed until market penetration, approximately in 2025. Especially municipalities can support market introduction by taking a role model function equipping public fleets with electric vehicles. Depending on market developments, this can be continued throughout all stages. Purchasing incentives are useful in the beginning of market introduction and should end before 2015. When e-vehicles are established in the market, one can also think of an incentive scheme that promotes environmentally friendly mobility. However, a representative survey (Pieper et al. 2013) revealed that vehicle tax exemption for 10 years is the more effective public incentive for the purchase of electric vehicles compared to free parking.

This section gave a brief theoretical overview of what politics can do and what, in the opinion of different authors, should be done to encourage the generation of innovations. There are some implications which can be drawn with reference to current activities to promote e-mobility. Firstly, e-mobility in Germany lags behind expectations, probably due to underinvestment and lack of linkages. Thus, funding programs have been set up which already had an effect on the number of built e-vehicles. As one can see in Fig. 1, there has been an enormous increase in the years 2008 and 2009 which has taken place simultaneously with the increase of public funding (Fig. 2). However, throughout the past three years numbers have decreased heavily again. Moreover, most of the vehicles are prototypes or concept cars and cannot be purchased at the market.

Since bringing forward e-mobility is considered important for public welfare, politics tries to increase investments as well as connections between different agents by promoting collaborative projects. This seems to be an adequate instrument according to literature. Nevertheless, the observed decrease in the amount of e-vehicles indicates that there is need for improvement. Whether this is due to government failure or other causes cannot clearly be stated. In order to avoid government failure, it is important to rely on the assessments of concerned agents regarding technological decisions, for example, rather than presuming knowledge Furthermore, bureaucracy has to be controlled and deadweight effects have to be prevented. The present study examines, inter alia, the impact of political measures on cooperation and innovation in the field of e-mobility and asks for the causes of poor innovative performance.

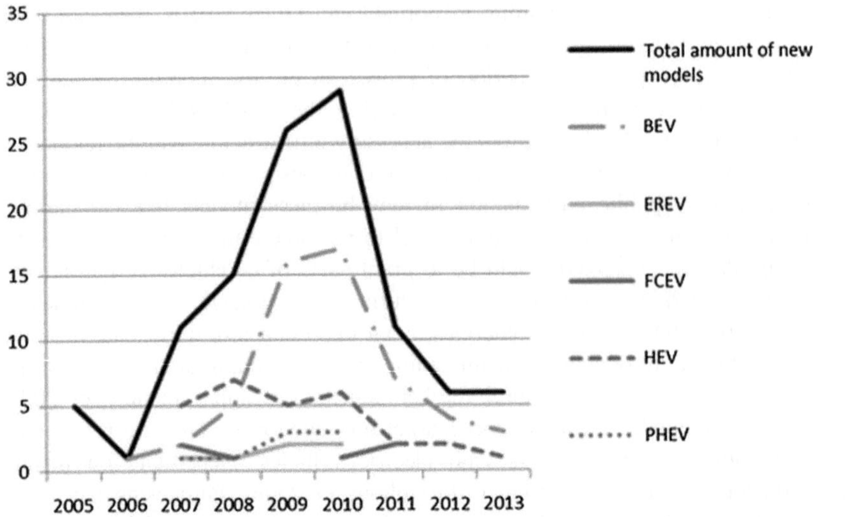


**Fig. 1** Amount of new models by German car manufacturers built from 2005–2013 (own database, as of 07/2013)

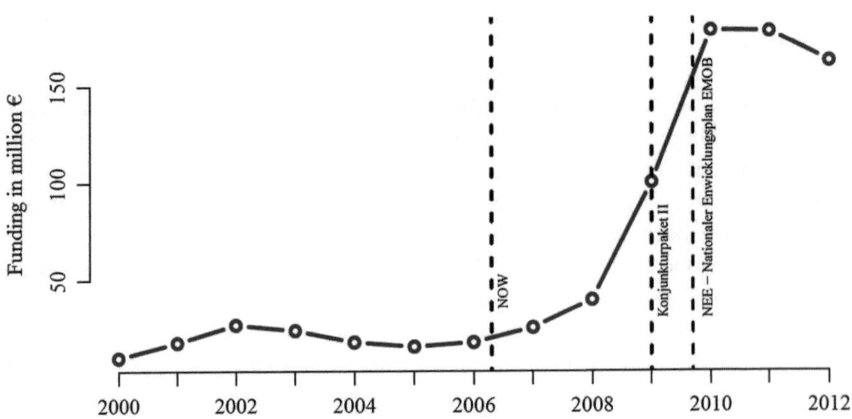


**Fig. 2** Development of public e-mobility funding in Germany

## 3 Methodology

Evaluating the interplay of cooperation and innovation has been the focus of the present research study. The presented views in this paper were gathered from a qualitative survey that was carried out in the context of a joint research project called "Neue Mobilität im ländlichen Raum". Besides a better understanding of cooperative behavior, the aim of the study was to identify impediments and potential for improvement regarding political measures. Altogether, 26 guideline-based face-to-face interviews have been conducted all over Germany, as well as one telephone interview. These include 23 interviews with experts from firms, universities and research centers that have been involved in innovations regarding different forms of electric mobility.[1] Selection criterion was the participation in collaborative projects funded by the EU or the German national government. In particular, it was tried to involve those organizations in the research study that have been able to gain notably large sums of public funding. This was based on a compilation of the five universities, research institutes and enterprises which attracted most public grants for each the national and European level. On the national level eight out of fifteen of the most successful organizations regarding the amount of funding could be interviewed, on the European level it was nine out of fifteen. Another three interviews were carried out with experts, such as lead partners for example, who can provide a more objective view on innovative development in e-mobility. A balanced selection of organizations of science and business has been involved in the survey. The exact allocation of participating organizations is depicted in Fig. 3.

Contacts were made by telephone whereas in many cases there was an exchange of information via email before the contacted persons agreed to take part in the survey. Ten persons and organizations, respectively, refused to take part in the survey or did not answer and could not be reached after first contact. Basically, the same guideline was applied in all interviews having been modified referring to the specific characteristics of the interview partner, however. On average, interviews lasted 46 min. They were carried out in German, except for one conversation that was held in English since the interviewee preferred to do it this way. Interviews have been recorded by means of a dictation machine and transcribed afterwards. Three interviews were scripted right after conversation, though, since the experts concerned did not give their permission to record the interview technically. After anonymization, a computer-assisted qualitative data analysis of the transcripts was carried out. Documents were explored and coded, i.e. sections of the texts were assigned to certain categories. In the first phase of exploration, open coding was carried out. This means that certain statements contained in the transcripts were summed up creating categories, thereby developing different codes. Text segments were consistently matched with the categories containing information about projects, innovation, cooperation, financing, politics and prospect. By means of text retrieval function information could be linked and interpreted regarding the specific context. This possibility of taking into account context-specific information and interpretive patterns also reflects one of the characteristic strengths of qualitative methods. With the help of the developed codes, variables were built allowing for an at least partly quantitative analysis, in addition. The survey focused on the cooperative behavior of varied organizations assuming that cooperation can enhance

[1]Electric vehicles in this context comprise all kinds of automobiles having an electric powertrain. This includes battery vehicles, hybrids and plug-in hybrids as well as fuel cell vehicles. Funding programmes, however, distinguish between the different sources of energy.

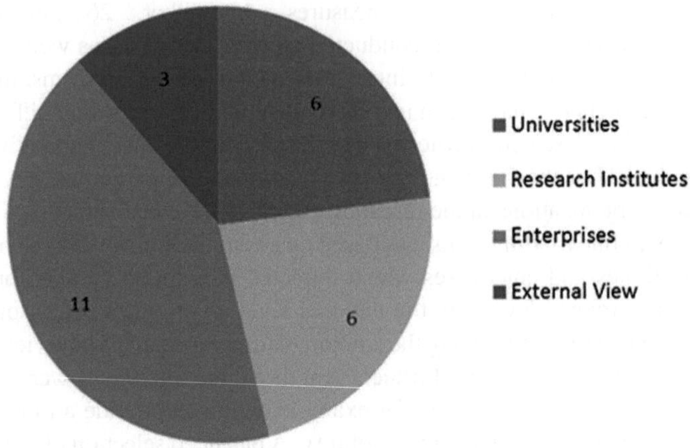


**Fig. 3** Allocation of interview partners regarding type of organization

innovation. However, the respective statements clearly indicate that political measures, especially in terms of public funding, have a strong impact on the success of collaborative projects and hence on the generation of innovations.

Compared to quantitative surveys, qualitative approaches can go more into depth and provide more openness for the persons addressed to articulate their personal views and opinions. Qualitative social research is characterized by a strong relatedness to subjective attitudes as well as by its tendency to incorporate different perspectives (Diekmann 2007; Flick et al. 2009). Cooperation and collaborative mechanisms are highly dependent on subjective attitudes and mindsets. These can hardly be ascertained by a standardized questionnaire. Therefore, deploying a qualitative framework is an adequate method for examining the presented research topic. Nevertheless, there are specific limitations regarding qualitative surveys. It is criticized, for example, that the interviewees are not part of a random sample, for example. Moreover, one has to be aware of general biases referring to personal interviews. In particular, that concerns the social desirability effect. Interviewees may answer questions in a way they think is appreciated by the researcher or answer questions regardless of not knowing an answer. Furthermore, results are not one-to-one applicable to the aggregate level. Replicability of content analysis is another common point of criticism (Diekmann 2007; Flick et al. 2009). Though, there are measures that can be taken to make analysis transparent. In the context of the present research study, it has been tried to make steps of analysis open to scrutiny by marking all relevant sections in the computer program, using memos in order to specify the applied codes and developing variables for a quantifiable evaluation of the data. Taking the objectives of the research study into account, however, guideline-based face-to-face interviews appear to be the most appropriate approach in order to explore the conditions and processes in collaborative projects.

# 4 Results of the Survey

The following section captures the statements of the interviewed experts and analyses them with regard to the influence of political measures. It was tried to show different strands of thought amongst them and interrelate those. Important aspects are highlighted using quotations to reflect the interviewees' arguments. In general, the interview partners share the opinion that policymakers are endeavored to support the expansion of electric mobility whereas efforts are not considered to have been very successful yet. About one half of the interviewees shares the opinion that policymakers are dedicated to support the shift towards e-mobility whereof eight interviewees even regard politics as initiator of the innovation system. However, some participants criticize that politics is either unwilling or unable to take steps in order to promote e-mobility. Following the view of all 26 experts, the most important instrument in this context is the allocation of public funding. The latter is so important that it was the most frequently mentioned topic throughout the interviews. As the interviews originally aimed at examining collaborative behavior, frequent reference is a strong indication for the impact of public funding on cooperation and innovation. Nevertheless, other aspects like personal contacts, spatial and social nearness highly influence cooperation, too, and will be examined in the following text.

Usually, public grants are given to collaborative projects. A lot of research projects have been funded this way generating new scientific and technological insights and enabling innovation. Most interviewed organizations have received pecuniary aids from the federal government. A little less prevalent is funding from the EU, projects financed by federal states or municipalities are even more seldom. However, there are many organizations which obtain funding allocated by different political levels at a time. In the context of this paper it is not always possible to distinguish between the different sources of public funding. However, the effects are considered to be very similar since measures follow almost identical principles and aims. Public funding can promote research and development in a field that would otherwise be too risky for certain organizations, thereby encouraging progress in a future technology with the potential to support a more sustainable form of mobility.

However, there are side effects caused by the dependence of all universities and most research institutes on project-oriented government grants. This implies that research can only be done on questions which form part of recently funded projects. 18 out of 26 interviewees stated that funding of specific core themes leads to research confined to respective questions. That means a restriction to scientific work and can be outlined as pretense of knowledge thereby indicating government failure. Besides, it can hinder innovations because it leaves little room for creative new ideas for which no funding program exists. This argument is also expressed in the following quotations[2]:

[2]Except for one interview held in English, all interviews have been carried out in German and have been translated for this paper.

> Scientific research is, of course, directly coupled with funding programs. So there won't be any scientific research in this field that is not linked to a funding program (040613_2, 17). Principally, now it is overhastily made up for lost time what you already had in mind 10, 15 years ago when there was no funding for that, though. [...] It is done now because now it is being funded, you already could have done it ten years ago (280513_2, 13).

Besides, public funding not only determines the focus of research and development interest, but its amount also affects the prevalence of collaborative behavior in innovation processes. Eleven of the interviewees find that funding capacity is good or even very good. Just one interviewed person, the employee of an OEM, totally disagrees and considers the amount to be just "peanuts" (230513, 27). One interviewee from a university also regards financial means not to be sufficient and further states that some project proposals would not receive acceptance regardless of good assessment due to low budgets. The generally positive estimation of funding amounts is formulated the most enthusiastic by a chemistry professor:

> If you work in the battery field Germany is the paradise now. There is no other country in the world like Germany for funding (160513, 64).

Money is also given to big enterprises which are not necessarily in need of funding to finance their R&D programs. This evokes criticism as some interviewees complain that innovation processes are slowed down by enterprises waiting for monetary aids despite having the means to do it without. Altogether, five interviewees suspect that the distribution of public money is partly politically motivated and that contacts to ministries and similar political institutions, respectively, are of high importance. They claim that small and medium enterprises had less opportunity to obtain funding than big enterprises, the latter already having enough money. Too, to their mind the spatial distribution of financial means is subject to political considerations. Moreover, some of them have the impression that good contacts to ministries are needed in order to be taken into account in the context of allocation decisions. Such criticism might be related to envy if other organizations are more successful in acquiring funding than oneself. Nevertheless, among those five organizations suspecting unfair allocation criteria are one big enterprise and one external expert organization, too. Furthermore, except for one research institution only enterprises declared to have valuable contacts to political institutions. One example for these connections is a big enterprise of which the interview partner reported that they were in close contact with the federal ministry of economics and could directly call the ministry in case of a new project idea. In contrast, the interviewee from a research institute that does not have established contacts to political institutions perceives this as a form of biased competition hardly providing transparency:

> You get the impression that it is rather a private circle (120613, 61).

Quantitative analysis[3] of funding determinants in the manufacturing sector also shows that certain characteristics of organizations affect their likelihood of being funded. There is, for example, a high degree of persistence in the funding system. The likelihood of obtaining funding is much higher if an organization has already taken part in a funded project before. This could be caused by increased know-how of administrative processes and the knowledge about the existence of funding programs in general. Degree centrality, i.e. the number of ties with other organizations, also increases the chances to be incorporated in the funding system. Besides, funding is especially attributed to young and big enterprises. Of course, from these analyses no direct conclusion can be drawn regarding the influence of political considerations on specific funding decisions. Moreover, it may be that there are branch-specific deviations from these findings. However, these analyses give a hint that there are aspects influencing allocation decisions that are not directly linked to the content of the proposals.

There is increasing competition for public funding between a plethora of organizations. Among those, there are also supposed to be some that have few experience in the relevant fields and just try to obtain as much funding as possible. The representative of an automotive supplier complains that there were professional recipients which were not able to do anything else than attract funding. A university professor summarizes the influence of public grants as follows:

> At times [you] almost got the impression that there is more money available than can really be spent usefully. And likewise institutions […], that […] simply jump on the bandwagon […]. In that respect you are almost in the situation of overfunding (220713, 75).

So there seem to be deadweight effects, i.e. organizations obtain funding without really needing it or being able to use it effectively—thus encouraging the waste of public monies. This further tightens competition. Competition, in turn, hampers cooperation as it leads to rivalry and more mistrust meaning that scientists or developers do not talk about their ideas or results too openly since they fear that somebody else might take advantage of that. The latter might imply that a competing research group adopts know-how using it to attract funding and leaving the other group without any financial means for research on the respective topic.

> The extremely competitive structure of funding is a cooperation obstacle, and a big one at that (280513, 3).

Moreover, volatility and policymakers' lack of perseverance in terms of changing research priorities are criticized, especially by interviewees from universities. If a technology does not make advances fast enough, another research focus will be funded which in turn will not be further pursued after a while. This is a prominent example for political decisions made under pretense of knowledge. In particular, this concerns the alternating concentration on either battery research or fuel cell research. It is assumed that it would be more effective if funding periods for

[3]These have been carried out by Florian Umlauf in the context of his Ph.D. research on cooperative determinants of research and innovation funding (not yet published).

certain research topics were longer and if there was parallel funding of both topics, respectively. Overall, that would save a lot of time due to the need to build up new research infrastructures and train scientific staff members every time again. These processes take up a lot of time, but as soon as scientific work has gathered pace it often does not take a long time until concentration again shifts to another focus. When after some years concentration is on the specific topic again all work has to be done completely anew as scientific staff has left the institutions, for example. In case of constant funding e-mobility would be much further ahead due to the opinion of many interview partners.

> I think that if you had started funding […] earlier and not that massively in shorter periods of time one certainly could have saved double labor or double funding (300513, 19).

Another example of lacking reliability mentioned by the interviewed experts is the funding commitment for the so-called show cases for electric mobility. Originally, it was planned to finance those by means of the energy and climate fund which should be supplied be proceeds from the $CO_2$ certificate trade. Due to fallen prices for $CO_2$ certificates, however, financing of these projects was uncertain for several months and it was not clear whether funding commitments could be kept. This led to a loss of trust:

> Trust has been damaged a little because we were left in abeyance for three, four months. Things like that are extremely harmful for such collaborative projects (040613_2, 89).

16 out of 26 interviewees criticize that application and allocation procedures are accompanied by high administrative efforts which bind a lot of resources and are particularly high in collaborative projects. Furthermore, it is criticized that procedures would extend over a too long period of time. Processes at EU level are considered to be especially difficult and complex. This even goes so far that the interviewee from a university states that they would not even try to acquire financial means from the EU as they had heard it was hardly possible. Moreover, one representative from a big enterprise states that collaboration between different federal ministries is not working well as they are behaving like rivals, in a sense. As evidence for that, he cites the fact that when applying for funds from a certain ministry, it is necessary to mention former projects positively that have been funded by this institution and not the ones funded by another.

The framework conditions which are connected with public funding particularly influence cooperative behavior. They especially affect the constitution of consortia and the selection of cooperation partners. These factors are regarded to be essential for successful cooperation by many interview partners. However, there are strict guidelines determining number, size, origin and type of participating organizations. Eight of the interviewed experts regard these guidelines as obstacles since these framework conditions would preclude or compel certain kinds of cooperation and increase the number of partners in a consortium too much, respectively. Therefore, in the opinion of some interviewees, the compulsory constitution does not always make sense:

> But that is the financially enforced allocation. And that is exactly what is not sensible in many cases (020713_2, 123).
> That cooperation which is as good as enforced by the fact that funding donors virtually always expect big teams, that to my experience is sometimes good and sometimes not so good […] because then cooperation runs as a hunting community rather than as contentual spirit and purpose (020713_2, 33).

Though, also improvement suggestions regarding the promotion of progress in e-mobility can be drawn from the interviews. The most commonly mentioned suggestions are the introduction of longer project durations as well as increasing the share of basic research funding. These suggestions are articulated by interviewees from universities as well as from enterprises. Among other things, this would offer the opportunity to create more fundamental innovations by enabling the involved people to think about issues anew. Some experts also recommend loosening framework conditions, first and foremost with respect to the selection of cooperation partners, since this promotes the building of trust and openness. A little bit more freedom is also suggested when it comes to the administrative obstacles related to the process of application. The latter often requires a very detailed project plan which leads to restraints by reducing the flexibility to react to unforeseen events. Instead, there should be a more careful examination of the results in the end for making sure that public money was spent responsibly. In addition to that, it would be helpful to reduce the time span of the approval process and to make binding promises in order to provide financial security. Besides financial aid for research and development, twelve experts argue in support of further political measures, especially subsidies for the purchase of electric vehicles, activities to raise the public's awareness for e-mobility and regulatory measures. The latter could, for example, imply free parking space for electric vehicles or the prohibition of access to urban areas with vehicles emitting carbon dioxide. Those suggestions are again drafted by diverse agents, but including all indirectly involved experts from external institutions.

## 5 Conclusion

A lot of efforts are made in order to promote the emergence and establishment of e-mobility in Germany. Intervention is considered to be necessary since there would only be insufficient investment without any incentives whereas it is intended to become lead market and lead supplier for electric vehicles. Insufficient investment especially concerns research and development activities, but also the investment in structural change in the industry which is quite risky. According to the theory of system failure (Metcalfe 2007), policies should support the self-organization of innovation systems by providing education, promoting bridging arrangements between different organizations and assigning credits.

Policymakers' commitment and endeavors are generally welcomed by the participants of the research study which per se is a positive signal. However, the conditions which determine the way in which agents cooperate have to be discussed. Not surprisingly, money seems to be the most important regulating screw to enhance innovative activities. Most interviewed experts regard the provided financial means as at least sufficient. This more or less generous spending has the side-effect, though, that there are many organizations competing for funding whereof some are not appropriate to receive monies as they do not have the means to use them for contributing to innovations. Nevertheless, they often form part of consortia since due to strict standards of number, origin and kind of partners, selection is very limited. So-enforced collaborations contain difficulties arising from lack of trust and social nearness, for example. The most important point of criticism, however, is the strong dependency of research on project-oriented funding of specific core themes. This dependency would lead to a restriction that might possibly inhibit innovation processes. Furthermore, tightened competition would make it harder for organizations to cooperate. Other important subjects of criticism were the modalities of placing projects and high administrative efforts.

It is a tightrope walk between the provision of enough financial means to trigger innovation on the one hand and the avoidance of waste of public funds on the other hand. Overall, the right instruments are chosen to support the innovation system, namely collaborative projects in order to promote exchange between different organizations.

Nevertheless, government failure can be observed to some extent. Deadweight effects can only be avoided partially. Thus, there is need to carefully evaluate project outcomes on a regular basis in order to identify organizations that do not use the provided means to create benefits. However, avoiding deadweight effects is quite challenging as it is very difficult to clearly determine which agents can contribute to progress in a certain field and which cannot, especially if consortia are very big. Furthermore, sometimes no quick progress can be achieved concerning particular research areas due to scientific or technological reasons like, e.g., experiments that are not working. The generation of transparency is essential to avoid the arousal of suspicions that policymakers would favor certain firms. Especially, transparency about the allocation of funding can be further extended to make sure that objective criteria are applied. There already are clear guidelines concerning application requirements and justifications are available in case of refusal. Therefore, information about funding programs, requirements and selection criteria have to be spread more widely and efforts to support small institutes or firms in overcoming administrative hurdles have to be intensified. Such measures might also help to distract reservations of potential applicants for public funding and augment transparency.

Some of the problems addressed here are not solely related to e-mobility-specific funding, but have deeper causes. This especially includes deficiencies in education and research policies, for example the underfinancing of universities. Project-based funding is often the only possibility to obtain money for research which restricts scientific work in terms of leaving no room to experiment freely. Besides, government presumes to know what the most appropriate line of research for the advancement of electric mobility is. This might lead to mistakes and delay.

Therefore, public funding has to be open and unbiased concerning technology. Research has to stay unaffected by statutory provisions in order to find out which solution is best for promoting e-mobility. Thus, it appears to be sensible to rethink and maybe restructure financing of public research in general as cutbacks of financial resources leave little room for project-independent research. Moreover, this would open up the opportunity to try out things and approach a task in new ways. To examine issues from different points of view may lead to more radical innovations and augment scientific knowledge constituting a value in itself.

Against the background of limited financial scope, the opportunity arises to shift the focus from pecuniary aids towards regulatory measures, too. A prominent example is the intention to ban conventionally-fuelled cars from European cities by the year 2050 as formulated in the European Commission's white paper (Europäische Kommission 2011). This seems to be a good possibility to leverage the automotive industry in order to step up efforts and to effectively change consumer behavior, thereby promoting alternative forms of mobility. However, it requires grit from policymakers to think outside the box and not only cling to interests of national industries.

## References

BMBF (2010) Ideen. Innovation. Wachstum. Hightech-Strategie 2020 für Deutschland. http://www.bmbf.de/pub/hts_2020.pdf

Bundesregierung (2011) Regierungsprogramm Elektromobilität. http://www.bmwi.de/DE/Mediathek/Publikationen/publikationen-archiv,did=390612.html

Czarnitzki D (2001) Die Auswirkungen der Forschungs- und Technologiepolitik auf die Innovationsaktivitäten ostdeutscher Unternehmen. Schmollers Jahrbuch 121(4):539–560

Diekmann A (2007) Empirische Sozialforschung. Grundlagen. Methoden. Anwendungen. 18. Aufl. Rowohlt, Reinbek bei Hamburg

Dudenhöffer F (2010) Elektromobilität: Deutschland verliert den Anschluss. In: Wirtschaftsdienst 90 (8), p. 502

Europäische Kommission (2011) Weißbuch. Fahrplan zu einem einheitlichen europäischen Verkehrsraum - Hin zu einem wettbewerbsorientierten und ressourcenschonenden Verkehrssystem

Europäische Kommission (2013) EU Wettbewerbsrecht. Wettbewerbsregeln für staatliche Beihilfen. http://ec.europa.eu/competition/state_aid/legislation/compilation/state_aid_01_09_13_de.pdf

Flick U, von Kardorff E, Steinke I (2009) Qualitative Forschung : ein Handbuch. Rowohlt-Taschenbuch-Verl, Reinbek bei Hamburg

Fornahl D, Diller M (2010) Phasenbezogene Förderung von Elektromobilität. In: Energiewirtschaftliche Tagesfragen 12, 2010, p. 70–76

Fritsch M (2011) Marktversagen und Wirtschaftspolitik: mikroökonomische Grundlagen staatlichen Handelns. Vahlen, München

JRC (2013) A look at the EU's research on electric cars. JRC News Release. https://ec.europa.eu/jrc/en/press-release/look-eus-research-electric-cars-9975?search

Kline SJ, Rosenberg N (1986) An overview of innovation. In: Landau R, Rosenberg N (eds) The positive sum strategy: harnessing technology for economic growth. National Academy Press, Washington, S 275–305

Maier G, Tödtling F, Trippl M (2006) Regional- und Stadtökonomik 2: Regionalentwicklung und Regionalpolitik. Springer, Wien
Metcalfe JS (1995) Technology systems and technology policy in an evolutionary framework. Camb J Econ 19:25–46
Metcalfe JS (2007) Policy for innovation. In: Hanusch H, Pyka A (eds) Elgar Companion to Neo-Schumpeterian Economics. Elgar, Cheltenham, S. 943–966
Mitchell WC, Simmons RT (1994) Beyond politics: markets, welfare, and the failure of bureaucracy. Westview Press, Boulder
Montmartin B, Massard N (2013) Is financial support for private R&D always justified? A discussion based on the literature on growth
Moreau F (2004) The role of the state in evolutionary economics. Camb J Econ 28:847–874
OECD (1997) National innovation systems. www.oecd.org/science/inno/2101733.pdf
Pieper N et al (2013) Elektromobilität auf dem Prüfstand - Wirksamkeit von Kaufanreizen. http://www.p3-group.com/downloads/2/1/0/6/7/Anschlussstudie_Elektromobilitaet_2013-06-10_ID_21065.pdf
Tokila A, Haapanen M, Ritsilä J (2008) Evaluation of investment subsidies: when is deadweight zero? Int Rev Appl Econ 22(5):585–600
Welsch J (2005) Innovationspolitik : eine problemorientierte Einführung. Gabler, Wiesbaden

## Author Biography

**Isa Reinecke** is research associate and doctoral student at the Centre for Regional and Innovation Economics (CRIE) at the University of Bremen. Her work is centered on entrepreneurship and networks. She holds a master's degree in urban and regional development.

# China's Leapfrog to New Electric Vehicles

**Peter Thomas in der Heiden**

**Abstract** In 2010, the Chinese government initiated a leapfrogging strategy to jump-start technological upgrading in the country's automobile industry and facilitate the transition to electric mobility based largely on indigenous innovation. This move was driven by escalating environmental pressures as well as resource constraints and inspired by the prospect of greatly strengthening the technological capabilities and competitive positions of domestic automobile manufacturers. The aim was to establish an early lead in the field of electric vehicles featuring high electric power ratios and power grid connectivity. The paper identifies this leapfrogging attempt as a strategic project initiated by the central government and organized in a top down manner. It sheds light on the supporting public policy framework and conducts a preliminary review of the strategy's accomplishments until mid-2012. By way of an in-depth examination of guidance policies and incentive structures, the paper highlights the challenge of coordinating promotion plans across the country.

**Keywords** Green innovation · Leapfrog development · Electric vehicles · China · Industrial policy

## 1 Introduction

In October 2010, the State Council (SC) released a groundbreaking document in which it designated seven strategic emerging industries (SEI) to be vigorously developed into leading pillar industries for future economic development:

P.T. in der Heiden (✉)
Mercator School of Management, Lehrstuhl Ostasienwirtschaft/China, Universität Duisburg, Essen, Germany
e-mail: petertidh@gmail.com

P.T. in der Heiden
Forsthausweg 2, 47057 Duisburg, Germany



D. Fornahl and M. Hülsmann (eds.), *Markets and Policy Measures in the Evolution of Electric Mobility*, Lecture Notes in Mobility,
DOI 10.1007/978-3-319-24229-3_7

environmental protection, new generation information technology, biotechnology, high-end equipment manufacturing, new energies, new materials and new energy vehicles (NEV) (SC 2010 at 2). This term was newly coined by Chinese authorities and includes battery electric vehicles[1] (BEV), fuel cell electric vehicles (FCEV) and plug-in hybrid electric vehicles[2] (PHEV). It thus comprises powertrain technologies with relatively high electric power ratios and power grid connectivity.[3]

Reiterating previous positions, China's cabinet outlined the need for growth to become more environmentally-friendly, resource-saving as well as technology driven. NEV development was to be attributed great importance by the automobile industry and other sectors. As opposed to the other six SEIs, NEVs represent a product ultimately intended for use by private individuals.[4] As the envisioned large scale adoption in this case hinges on the individual perceptions and behaviour of several million car buyers, promotion policies had to reach down to private consumers. However, insufficient resources and institutional constraints on the part of central ministries required that local authorities serve as vital link between central government and individual consumers.

## 1.1 The Dilemma of Policy Implementation in the Age of Economic Policy Decentralization

Decentralization of decision-making power has evolved into a major theme in the literature on Chinese economic development, especially in the post-Mao period. Over the course of successive reform steps, economic management authority has gradually shifted to regional governments. Oi (1998) has coined the term *local state corporatism* reflecting the idea that leading officials at and below the municipal level may not only have close ties to local enterprises but supervise economic

[1]Battery-electric vehicles solely rely on electricity for propulsion. In order to replenish its charge, the vehicle has to be connected to the power grid. Through advanced communication protocols, the vehicle can send and receive information through this link. The powertrain design is fairly straight forward, with a battery powering an electric engine.

[2]Plug-in hybrid electric vehicles are designed with two power sources for propulsion. A gasoline-powered engine may either directly transfer kinetic energy to the wheels (parallel configuration) or charge a battery that powers an electric engine to provide propulsion (serial configuration). Additionally, the on-board battery of plug-in hybrid electric vehicles can be connected to the power grid to recharge and exchange information.

[3]Electric vehicle is a very broad term that captures all types of powertrain architectures which can make use of electricity stored in on-board batteries for propulsion. EVs include both hybrid electric vehicles (CHEVs and PHEVs) as well as pure-electric vehicles (BEVs and FCEVs). Following the classification used by the Chinese government, EVs can be equally understood as including both NEVs (BEVs, PHEVs and FCEVs) as well as CHEVs.

[4]Unless explicitly stated otherwise, this article is exclusively dealing with passenger cars of different powertrain architectures. Commercial vehicles and other types of passenger vehicles were not part of the analysis.

activities—effectively acting as board of directors. Furthermore, local authorities are uniquely placed to take advantages of multiple sources of finance—mainly through extra-budgetary revenue—and enjoy greater control over fiscal resources. Zheng (2007) introduced the concept of de facto federalism, arguing that in China's hierarchical political system, tasks are divided between central and regional governments, giving both players more or less mutually exclusive domains in which they can make decisions. He maintains that intergovernmental decentralization has been institutionalized in ways that cannot easily be undone. This has made it increasingly difficult for the centre to infringe upon policy areas which are the domain of regional governments.

The fierce competition between localities for corporate investments and infrastructure projects has also produced a large variety of development strategies and policy outcomes—a phenomenon (Howell 2006) labelled a polymorphous state. He perceived the Chinese state as assuming "multiple complex forms and behaviours across time and space, defying its reduction to a unitary actor" (Howell 2006, p. 273). On this backdrop, Howell argues, policymaking undertaken by central authorities and targeting economic development on a national scale, is undermined by localities pursuing parochial and often conflicting profit maximization strategies. Economic policy decentralization has thus put national policy strategies at odds with a large number of local development states, causing sub-optimal policy outcomes and weakening the ability to direct national economic development.

This dilemma is closely related to the contentious relationship between two coexisting administrative frameworks: the tiao (branches) and kuai (pieces) systems. While both are used simultaneously, they possess several mutually incompatible features (Unger 1987). The tiao framework is built on the hierarchical structure of functional administrations, e.g. finance departments, penetrating all administrative levels. Central ministries and commissions are in charge of the planning, financing and supervision of policies based on decisions reached by senior leaders. Instructions are channelled downward and outward to functional departments at provincial, prefectural and township levels. While the branch framework allows for the rule of experts and promotes efficient coordination at the national scale, it is prone to local coordination failures as it may not sufficiently take into account the opportunities and needs of particular localities. The kuai framework, on the other hand, has party committees of regional jurisdictions as its main actors. Policies and instructions are handed down through the party organization, thereby assigning more power to generalist cadres. While the kuai framework enables localities to engage in comprehensive economic development planning, it is prone to cause duplicative investments and coordination failures across regions.

The local leadership receives policies from higher level functional departments through the various branches. It adjusts those instructions taking into account local conditions and coordinates them with the localities' very own development strategies. Functional departments on the local level are subject to the dual authority of their local leadership as well as their functional superiors on the next-higher administrative level. Problems arise once directions passed down from functional superiors at the central level contradict local objectives or entail high implementation costs.

Consequently, the key question is how it can be ensured that central policy is effectively implemented locally—even in the face of conflicting economic preferences.

Lieberthal (1992) argues that under the conventional authoritarian political system which relies on coercion exercised through rigid command structures, local compliance could only be ensured if firstly, the central leadership is united in their endorsement for a particular policy; secondly, the central government assigns priority to the policy in question and provides a coherent action plan and thirdly, the implementation performance on the local level is measureable. Whenever these three conditions are met, policy implementation can be expected, as non-compliance would be sanctioned. However, Lieberthal (1992) concedes that such cases are few and far between. According to Lieberthal and Oksenberg (1988, p. 3), the fragmented, segmented and stratified nature of the Chinese policymaking process mandates "a system of negotiations, bargaining and the seeking of consensus among affected bureaucracies" implying that any major policy would have to win support and cooperation from a host of agencies.

Lampton (1992) posits that the need for inter-governmental bargaining and negotiations has been an established practice long before the onset of the reform era but assumed greater relevance after 1978 because the complexity and inter-dependency of development agendas of various sub-central governments have increased substantially. He maintains that bargaining processes are time consuming since striking a consensus between multiple players with diverging objectives is a complicated endeavour. Time constraints may render it unpractical to seek definite decisions on all relevant aspects all at once. Thus, initial agreements are likely to be found in principle only, saving up further negotiations on detail issues for the implementation phase. Furthermore, he maintains that in order to obtain policy implementation through negotiations with sub-national governments, central authorities have to first establish internal agreement on critical issues and be ready to devote substantial political and economic resources to the negotiation process.

Based on these previous studies, the author argues that conflicting agendas of central and regional governments as well as collusive behaviour of local governments and weak supervision capabilities of central authorities have stalled the ambitious development strategy for NEVs. The paper seeks to combine the two broad themes of central-local relationships and technical leapfrogging in order to interpret the outcome of the central government's strategy to promote NEV development.

The present analysis draws on a series of 17 interviews with leading experts in companies, international organizations, industry associations and universities conducted in the first half of 2012 in China. Furthermore, it bases on a comprehensive review of official documents issued by government departments with various functional responsibilities and on different levels of the administrative hierarchy.

To the best of our knowledge, the present analysis represents the first paper conducting an in-depth analysis of the coordination, cooperation and conflicts between central and local government organizations in the sphere of SEI in general and NEVs in particular. It is part of a larger research project dedicated to exploring the broader concept of government failure as originating from unsuccessful attempts to remedy market failure and selective industry promotion.

## 1.2 Methodology

The methodical approach for this paper rests on two pillars, desk research and expert interviews. The first consists of the in depth study of government sources by means of desk research. The author enjoyed an extensive access to government issued regulatory, planning and assessment documents as well as reports and statistics. The review of documents from a wide range of government organizations stretching from the state council and central level ministries and commissions to the planning and functional departments on the municipal level has proven an invaluable source of information. Complemented by Chinese language articles in academic journals and on news websites the material obtained was complete enough to render own quantitative studies unnecessary.

For the second pillar, the study employed purposive sampling to identify industry experts to inform the study (Emmel 2013). Interview partners were chosen on the grounds that they possessed deep insight and profound knowledge of electric vehicle development and commercialization in China. Following a contact request sent to 73 research institutes, automobile companies, suppliers, industry associations and other organizations, 20 individuals agreed to be interviewed either face-to-face or by phone. The sampling draws and expands on a list of distinguished industry experts contained in the 2012 study on the competitiveness of the European Automobile Industry in the Field of Electric Vehicles conducted by Proff et al. (2012) and commissioned by the European Commission. In order to reduce bias and obtain a more comprehensive set of information, interview partners were deliberately selected from a rather wide spectrum of activities (Flick 2014). The sample consists of senior managers, administrators and academics—all with an in-depth understanding of the development dynamics, industrial policy concepts and outcomes in the electric vehicle field.

Interviews were qualitative and semi-structured. They covered three main content areas. (1) The current situation and future expectations for the Chinese EV industry. This includes questions on competitive strength and weaknesses of Chinese companies as well as the opportunities and threats they confront. (2) The characteristics of industrial policy measures by government organizations on various levels and of different functions. This includes questions concerning the effectiveness and efficiency of such measures. (3) The dynamics of inter-governmental and state-business relationships. In this context, participants were also asked about their respective organization's involvement in the policy-making process as well as the impact government regulation had on them.

The diverse professional backgrounds and organizational affiliation of informants made the use of a standardized survey questionnaire unpractical (Babbie 2010). Consequently, the interviews followed a common protocol but allowed for putting topical focus on areas where subjects were the most knowledgeable. As all interview partners agreed to participate under the condition that their responses remained confidential, many volunteered information and contributed significantly to the data gathering effort.

Conducted in 2012 and 2013 in both China and Germany, interviews took between 1 and 3 h. Until the end of 2013, follow-up interviews were held over the phone to obtain clarifications on their earlier statements and discuss new developments as information became available. In a second step, a thematical content analysis was undertaken to structure and analyze the statements in order to identify consensus groups in topical areas. In a third step, the various conceptions of policy and corporate strategies as well as inter-governmental and state-business relationships were related to information obtained through desk research.

## *1.3 The Leapfrogging Concept*

Technological leapfrogging (TL) was introduced by Soete (1985) in the context of the international diffusion of technologies and has since been subject to much debate. Steinmüller (2001, p. 194) defines the phenomenon as "bypassing stages in capacity building or investment through which countries were previously required to pass." It is important to distinguish TL from catch-up development. The former implies that latecomers pursue advanced competitors by improving and accelerating their technological development. Following the path taken by forerunners and imitating their products and processes, latecomers may gradually close the development gap. TL, however, implies that latecomers not only move faster than industry leaders, but do so in non-continuous ways. Leaving out development stages allows latecomers to move ahead of forerunners and claim technology leadership (Chen and Li-Hua 2011).

Sood and Tallis (2005) argue that the early development stages of a new technology paradigm present the best opportunities for realizing TL. This is due to the fact that no player has established a dominant position and forerunner-latecomer roles have not yet been assigned. Naturally, a situation in which all competitors set out from the same starting line offers more favourable conditions than one in which a latecomer has to struggle first to close a development gap and then to surpass the industry leader. However, early entry is very challenging since there are no development blueprints to imitate and no established pathways to follow. Therefore, autonomous innovation and independent R&D are the only tools available.

Fu and Zhang (2011) maintain that obtaining cutting edge technology from outside sources is costly and difficult. Leading companies are highly protective of core competences, as these sustain their competitive advantage and are reluctant to share crucial technologies with potential rivals. Furthermore, the elements of new technologies that can be purchased are limited to explicit knowledge. Tacit knowledge, which is essential for technology mastery is usually not available. Moreover, latecomers need to possess the absorptive capacity to make the best use of imported or acquired technology. Without strong own R&D capabilities, technology assimilation is impossible. It may also be the case that transfers yield limited rewards as the underlying technology may be unsuitable for the receiving country (Fu and Zhang 2011). Consequently, indigenous innovation represents a conditio-sine-qua-non for successful TL. Chen and Li-Hua (2011, p. 95) argue that "TL in technique and products can only be realized when an enterprise adopts self innovation and improves its independent technological innovation capacity greatly enough to catch-up with or surpass the leading corporations in its industry." Similarly, Yuan (2002) emphasizes the crucial role of independent research and innovation as well as proprietary technologies.

With regard to technology diffusion patterns prevalent in China's automobile industry, most experts agreed that substantial technology spill-overs have occurred between overseas and Chinese OEMs, mainly through the government-mandated pairing in Sino-foreign joint ventures. However, asked to evaluate autonomous R&D capabilities and innovation capacities, experts maintained that only very few Chinese players had reached levels comparable to global OEMs. Most companies had either neglected autonomous R&D in the face of close cooperation with foreign partners or simply failed to muster the financial and human resources necessary for R&D in cutting edge technologies. Even under institutional arrangements specifically designed to facilitate technology transfers from overseas OEMs in exchange for market access, Chinese peers were not able to close the development gap separating them from their significantly more advanced foreign partners. Technology diffusion was sluggish due to the five obstacles Fu and Zhang had highlighted. On the background of past experience, the Chinese government has placed special emphasis on indigenous innovation in the sphere of EVs.

Fu and Zhang (2011) as well as Chen and Li-Hua (2011) assign great importance to the role of government authorities in creating favourable institutional environments and guiding indigenous innovation. While Fu and Zhang suggest that the state should initiate the transition and maintain its momentum, Chen and Li-Hua maintain that a national technology strategy is conducive to TL. This paper argues that the Chinese national EV strategy was launched and directed by government authorities.

Another important theoretical aspect deals with the question if TL can happen in developing countries. Conventional wisdom suggests that significant technology leaps only take place in developed economies which already host high-tech industries and can play out advantages such as advanced R&D infrastructures, relatively complete institutional settings, high education levels or large capital stocks. For most of the time, technological change progresses in a gradual fashion based on incremental improvements. This "normal" process is based on learning by doing and mainly takes place in developed economies. From time to time, R&D breakthroughs bring about disruptive technologies which rock the foundations of the incumbent technology paradigm. Schumpeter (1947) proposed that leading firms face fewer incentives to innovate than lagging ones and will thus eventually lose their frontrunner status as followers jump ahead using new and superior technologies. Commitments to previous technology generations constitute legacy costs which make TL a less attractive strategic option. As the transition to a new technological paradigm would require substantial upfront investment in R&D, plant equipment or employee training, incumbents may prefer to hold onto established technologies. Less advanced players may thus claim an early lead. Brezis et al. (1993) introduce international differences in labour costs into this concept and argue that leading countries have higher wages than lagging countries. Because at the outset new technologies are less productive than old ones, their adoption appeals more to low-wage countries. Using a model with two goods and two countries, they explain why previous success may breed future failure.

This notion is highly relevant to the case of leapfrogging to EVs in China, not only because the country features lower average wages than the home countries of the world's leading automobile companies. Another important issue is that Chinese firms, due to their relatively short modern development history, have not yet been able to put down as much investment in established powertrain technologies as their more advanced rivals. With relatively little proprietary technology and a relatively big development gap separating them from overseas competitors in the field of internal combustion engine (ICE)[5] and conventional hybrid electric vehicle[6] (CHEV) technologies, Chinese OEMs may be uniquely placed to perform TL.

## 2 The Rationale for Technology Leapfrogging and Supporting Policies

The pursuit of TL in the automobile industry represents a component of national industrial policy framework which promotes R&D and technology upgrading in order to enhance national competitiveness and prevent China from getting stuck in the middle-income trap. The domestic automobile industry, which benefitted substantially from the prolonged car boom in its home market and technology spill-overs from foreign joint venture partners, still lags behind foreign peers in terms of R&D capabilities and innovation capacity (Meier 2009). With U.S., European and Japanese automobile manufacturers all commanding a solid lead in ICE technologies, the prospects for Chinese players to catch up look dim. Similar things can be said about the CHEV field, where Toyota had launched volume

[5]Internal combustion engines convert liquid fossil- or bio-fuels into kinetic energy and represent the most commonly used propulsion technology for road vehicles in the past WWII era.

[6]Conventional hybrids are characterized as being equipped with both an internal combustion engine as well as an electric engine. On-board batteries have a very low capacity and make negligible contribution to vehicle propulsion. CHEVs cannot be connected to the power grid but are recharged by means of recuperative breaking and related technologies.

production of its Prius as early as 1997. Following this line of argument, NEVs are characterized as infant industry to be nurtured and protected so it may develop into an important pillar of the future economic structure (Proff et al. 2012).

This situation gave rise to the notion that the Chinese automobile industry should put special emphasis on the development of green powertrain architectures that promised the largest long-term potential. Both BEVs and PHEVs were considered as cutting edge technologies but also as achievable targets considering that firstly, key technologies were still in their early development stages. No automobile manufacturer anywhere in the world had claimed a technological breakthrough that translated into volume production and no individual player could claim technological leadership (Proff et al. 2012). Secondly, Chinese automobile manufacturers had not yet amassed legacy costs related to previous powertrain technologies in the same extent as their overseas competitors. Adjusting the focus of development and production should thus have been relatively easy. Thirdly, key raw materials, mainly Lithium and rare earth elements, which are necessary inputs for the manufacture of core components, are conveniently available from domestic mining and production. Fourthly, the previous expertise in the mass-production of batteries and existing capacities in that industry represented competitive advantages. Fifthly, based on the conviction that resource bottlenecks, environmental crisis and technological progress would make the electric mobility an inescapable long-term trend for any country, the large volume and high growth of its domestic market presented favourable conditions for launching NEVs (in der Heiden and Taube 2012).

The decision of the SC (2010) to prioritize and accelerate development in selected areas was not the first major attempt to stimulate the EV market. As early as 1999, fourteen central ministries and commissions had launched a joint initiative to promote clean vehicles, including but not limited to EVs. Chinese authorities hoped to kick-start the popularization of EVs in order to curb air pollution and directed the automobile industry to engage in related R&D (MOST et al. 1999). This endeavour, however, failed to make a major impact on the structure of new vehicle sales. While consecutive steps to tighten fuel efficiency and exhaust emission standards in the following years succeeded in substantially improving the environmental performance of cars, little progress was made in terms of putting more EVs on Chinese roads. Nevertheless, the central government continued to emphasize the importance of developing and popularizing EVs in its plans and programs. The *Automobile Industry Policy* of 2004, for example, called for the active development of such vehicles (NDRC 2004 at 3.8). Similarly, the *Adjustment and Revitalization Program for the Automobile Industry*, released in 2009, addressed the issue of EVs. After restating earlier calls for redoubling R&D efforts and launching marketable products, the Program went beyond the scope of previous documents by specifying concrete development targets and announcing the provision of support funds. Until the end of 2011, the Program ruled, automobile manufacturers should establish production capacities for 500,000 EVs (SC 2009 at 2.3.6). Furthermore, EVs should represent 5 % of total new vehicle sales by that year. In order to achieve these aims, the document pledged RMB 10 billion worth of subsidies, allocated through a special fund under the central budget, to R&D

projects including but not limited to EVs (SC 2009 at 4.9). Until May 2010, all documents published and initiatives launched with the intention of promoting vehicle electrification shared one feature: they did not prioritize any particular type of powertrain architecture but indicated government support to a broadly defined group of advanced propulsion technologies conforming to the twin objective of resource conservation and environmental protection.

The decision of the SC to designate NEVs—a subgroup of EVs—as SEI marked a striking departure from past practice and had profound influence on the orientation and intensity of future promotion policy. According to the Ministry of Science and Technology (MOST), "Confronted with the grim challenge of conserving energy and reducing emissions, cultivating NEVs as SEI and realizing indigenous innovation as well as the historical task of technological leapfrogging, the development of NEVs has already evolved as a major national issue of strategic demand and strategic focus." (MOST 2012 at 1.3). Consequently, when the National People's Congress (NPC) approved the *12th National Five-Year Program* in mid-March 2011, the section dedicated to outlining the government's vision for the SEIs was titled "Promote Key Areas' Leapfrog Development" (NPC 2011 at 3.10.1).

# 3 Promotion Policies

The government pledged to formulate standards, organize demonstration projects and facilitate the application of major innovations. More concretely, central authorities announced the initiation of a special fund to promote R&D activities and channel social investments into SEIs. In order to increase access to vital investment capital, financial institutions were urged to step up loan support. To this end, preferential financial policies, such as risk compensation schemes, were to be put in place. Taxation policies were to be amended to encourage innovation and stimulate investment as well as consumption in the NEV industry and the six other SEIs (SC 2010).

a. Demonstration projects

According to the retardation hypothesis (Soete 1985), adoption of a new technology by potential users is initially held back by a lack of relevant information and uncertainty concerning its specific performance parameters. By observing early adopters and learning from their experience, potential users obtain a better understanding of the new technology causing the level of uncertainty to drop. If the new technology can meet users' requirements, this may lead to imitative buying behaviour that may cumulate in a bandwagon effect.

In line with this argument, the Chinese government attributes high priority to large-scale demonstration projects intended to showcase the advantages of EVs. Besides helping manufacturers increase their output numbers and sales revenues, demonstration projects also allow for long-term testing of EVs in significant numbers and under real world conditions. Of additional importance is the fact that demonstration projects increase the visibility of EVs in the public domain and allow people to familiarize themselves with the new technology (Proff et al. 2012).

In early 2009, the MOST and the Ministry of Finance (MOF) initiated the *10 Cities, 1000 Vehicles Program*,[7] supporting 25 municipal governments from across China to adopt EVs into their fleets (MOF and MOST 2009 at 1.2). Since this particular scheme was launched before the strategic focus was placed on NEVs, subsidization is not limited to this subgroup of EVs. This program is reviewed here because as the largest EV demonstration project launched so far, it has strong implications on the development and popularization of NEVs. Furthermore, it illustrates the preferences and procurement practices of municipal governments, shedding light on serious coordination problems.

Every participating city had to commit itself to launching a certain number of EVs and constructing related charging and maintenance infrastructures until the end of 2012. The vehicles were to be used by government departments and affiliated units as official cars, taxis, public buses, sanitation or postal vehicles. While actual purchase costs were covered by central government subsidies (see Table 2 in the appendix for an overview), municipal authorities had to invest their own money to maintain the vehicles and set up supporting infrastructure (MOF and MOST 2009).

cAs units under the ultimate supervision and authority of the central government, municipal governments were entrusted to administer pilot projects within their respective jurisdictions. Furthermore, the initial lack of charging infrastructure necessitated a focus on fleet operators in areas like public transport, sanitation or postal services to take advantage of predictable driving patterns (World Bank 2011).

In May 2010, the Chinese government launched a separate pilot program to promote private sales of NEVs (MOF et al. 2010). Beijing, Shanghai, Changchun, Hangzhou, Hefei and Shenzhen were selected to participate in this scheme which was scheduled to conclude by yearend 2012. Pursuant to the program's regulations, purchasing premiums were awarded exclusively to private buyers of PHEVs and BEVs with battery capacities of no less than 10 and 15 kWh respectively.[8] In both cases, every kWh of battery capacity in excess of the minimum requirement was rewarded with subsidies in the amount of RMB 3,000—up to a maximum of RMB 50,000 for PHEVs and RMB 60,000 for BEVs. Municipal governments were tasked to administer the program and take responsibility for the installation of charging infrastructure. Furthermore, they were ordered to offer discounts on electricity rates for vehicle charging, establish special parking lots reserved for

[7]The title of the Program is misleading in two regards. The initial number of 13 participating cities was expanded to the current 25 before the end of 2009. Furthermore, adoption targets were not uniformly set at 1,000 units but differed substantially according to local conditions. Beijing, for example committed itself to launching over 5,000 units while several smaller cities pledged to introduce little more than 1,100 units.

[8]Vehicles using lead-acid batteries are disqualified. Other requirements include that BEVs can travel a distance of at least 50 km on a single charge, are able to maintain top speed for at least 30 min and meet a large number of other criteria pertaining to speed, acceleration, energy efficiency, slope climbing and other parameters.

NEVs and—most importantly—to implement their own subsidy schemes. In compliance with the last demand, the six participating cities have crafted very different regulations (see Table 1 in the appendix for an overview).

b. Contribution of local governments

Although the central government has been the driving force behind EV promotion, the fact that successful large-scale adoption ultimately depends on the buying decisions of ordinary people made it clear from the very beginning that local governments would have to play a major role. Firstly, the financial burden stemming from the provision of R&D subsidies, purchasing premiums, tax incentives and other promotional measures was considered too large for the central government to shoulder. Secondly, cooperation of lower level authorities was necessary for the implementation of demonstration projects "on the ground", as central authorities lack the institutional infrastructure and manpower to carry out such plans throughout the country. Thirdly, coordination with local government was essential for managing non-monetary incentives. Exemptions for NEVs from buying, driving and parking restrictions are administered on the local level. The same is true for discounts on parking, maintenance or highway fees as well as privileged access to parking lots or driving lanes.

As all major initiatives to stimulate EV sales hinge upon the active participation of local governments, in late 2011, the SC ruled that all jurisdictions should carry out low carbon pilot projects "in light of their actual circumstances" (SC 2011 at 3.9). Local authorities were urged to proactively devise low carbon development plans and commit themselves to green consumption. Model projects should give priority to the development of EVs and other low carbon technologies and expand over time (SC 2011).

## 4 Current Market Situation

In mid-2012, 13 years after the development and popularization of EVs was first mentioned as a priority task in the *Air Purification-Clear Vehicle Plan,* China's EV population reached 16,000 units (China.org.cn 2012). Much of that number had been added during 2011, as a total of 8,159 EVs (5,579 pure-electric vehicles[9] and 2,580 hybrid electric vehicles[10]) were registered during that year (Xinhua 2012a). Relative to total Chinese motor vehicle sales in 2011 of 18.5 million units, EVs—all types of electric powertrain architectures taken together—accounted for just 0.4 %.[11] This result is a far cry from the 5 % market share target set for this year by

[9]Pure electric vehicles include BEVs and FCEVs.

[10]Hybrid electric vehicles include PHEVs and CHEVs.

[11]It should be pointed out that the general sales numbers do not contain purchases made by government agencies under the 10 Cities 1000 Vehicles Programme.

the Adjustment and Revitalization Program for the Automobile Industry, cited earlier.

While the Chinese EV market was dominated by domestic OEMs until 2011, this situation has changed dramatically in 2012, as Japanese brands have assumed a more prominent role. Even the row over the Diaoyu Islands did not interrupt this trend, as Toyota-branded vehicles claimed roughly two thirds of the market for EVs in August. The combination of steep sales growth of the Prius (a CHEV manufactured in China) and the successful introduction of the Camry Zunrui (a CHEV imported from Japan) in that month, the Chinese EV market more than doubled to 1,065 units from previous month (EVTimes 2012).

By the end of 2011, the overall progress of the 25 cities participating in the *10 Cities 1000 Vehicles Program* was modest. Two and a half years into the Program, only 11,949 out of the targeted 52,621 EVs had been obtained by city governments—accounting for just 22.7 % of the adoption objectives (Xinhua 2012b). Only four cities could present a plan fulfilment in excess of 30 % (Hangzhou, Zhengzhou, Suzhou and Beijing). At the lower end of the ranking, Tangshan stood out for its low goal attainment of just 2 %. In terms of absolute numbers, only five cities had introduced more than 1,100 EVs with Shenzhen taking a lead with 2,571 units (AutoQQ 2012). Conversely, at the end of the league, Xiangyang had launched a meagre 7 EVs. In terms of vehicle type, most of the EVs obtained by city governments and affiliated businesses were buses (77.6 %) while passenger car purchases were few and far between. With regards to propulsion systems, cities had shown a striking preference for hybrid electric vehicles while shunning pure-electric ones (Xinhua 2012b). In terms of the establishment of supporting infrastructure, which had also been covered by the Program, overall progress had been lagging expectations. Jinan, for example has only constructed a single combined charging and swapping station—which reportedly is too small for buses to turn around (21st China Business Herald 2012a, b)—while Chongqing has installed just 150 charging poles (Xinhua 2012b).

Similar to the case of local government and public service units, the promotion of private NEV sales in six pilot cities has yielded disappointing results as well. At the end of 2012, 18 months into the project and 12 months before its scheduled end, the overall goal fulfilment of the six pilot cities stood at about 2 % (Fourin 2012). Hefei and Shenzhen stood out as either city had registered cumulative sales in excess of 600 units indicating fulfilment rates of 3–5 % (21st Century Business Herald 2012a).

As the progress of the two major demonstration projects has illustrated, the introduction of electric passenger cars has been exceptionally slow compared to other vehicle types. In the *10 Cities 1000 Vehicles Program* which targets municipal governments, purchases of cars have lagged behind buses and in some cases even sanitation trucks. A preliminary review of the achievements of the scheme for promoting private NEV purchases confirms that purchasing incentives have so far failed to convince auto buyers to choose electric cars over conventional ones.

## 5 Evaluation of Coordination Failures

a. Coordination failures on the central level

The initiation document of the SEI strategy emphasized comprehensive planning and guidance. By way of a national development plan for SEIs in general, special plans for each SEI as well as the integration with other relevant plans and policies, it was hoped to maintain a tight grip on the development process. The regular collection of statistical data and compilation of progress reports was intended to facilitate effective supervision and direction. Furthermore, strong guidance of the local governments' development strategies was announced so as to ensure an optimal regional layout and the formation of comparative advantages through coordinated efforts. Even at this early stage, the SC warned that blind development and redundant construction could undermine progress (SC 2010 at 8.3). As this chapter will illustrate, these standards have not been adhered to as planned.

On the central level, the departments directing NEV promotion, primarily the NDRC, the Ministry of Industry and Information Technology (MIIT), the MOF and the MOST to some extent hold diverging opinions. According to the industry experts consulted in early 2012, participating agencies struggled to establish consensus concerning the amount of support funds and their distribution across different powertrain platforms as well as other contentious issues. While the SC document initiating the leapfrogging strategy had promised to strengthen inter-agency organization and "establish an interdepartmental coordination mechanism for the development of SEIs led by the NDRC" (SC 2010 at 8.3), this did not happen. As the division of responsibilities was subject to internal debate, the policy-making process was much less stringent than had been initially envisioned.

During the interviews, experts pointed out that the delayed release of the *Development Program for the Energy-saving and New-energy Vehicle Industry* (2012–2020) had been a cause of uncertainty for companies and thus undermined NEV development. The Development Program, of which an initial draft for solicitation of opinions was released in October 2010, had been the subject of intense debate among the ministries and commissions involved in the policy-making process. After the SC had rejected two draft versions, the final document was eventually published in July 2012 (SC 2012). It continued to assign top priority to NEVs. However, the slow progress of technology development, infrastructure expansion and consumer acceptance in this field did not escape the authors' attention. Furthermore, central government departments observed the EV market took an unwanted turn towards the less favoured CHEVs. These vehicles were supplied by Sino-foreign joint ventures or foreign companies and contained little, if any, proprietary Chinese technology.

In light of these developments, CHEVs were designated as the dominating electric powertrain technology until 2020. However, the Development Program stopped short of abandoning the previous commitment to TL as it aims for NEV commercialization after 2020 (see Appendix 2 for the official development timetable). During the interviews conducted in early 2012—before the release of the

Development Program—the central government's leapfrogging strategy had been met with scepticism by experts, most of whom warned that this approach was inconsistent with consumer preferences and unrealistic considering technological capabilities in underlying core technologies, such as traction batteries, power electronics and electric engines.

Since the structure of purchase premiums, R&D subsidies and other promotion measures was strongly tilted in favour of NEVs, CHEVs had not assumed a high profile with Chinese auto makers. Classified as energy-saving vehicles, purchasing premiums were limited to RMB 3,000—the same amount offered for efficient gasoline powered vehicles (MOF, NDRC and MIIT 2011). Consequently, Chinese OEMs focussed their efforts on NEVs while neglecting CHEVs. The Jiexun, independently developed by Chang'an Motors, and launched in March 2007 was China's first completely home-grown CHEV. However, disappointing sales of only 1,300 units over a period of 30 months prompted the manufacturer to discontinue the vehicle in late 2009 (AutoQQ 2010). After this failed attempt and the advent of TL policies, Chinese companies had readily moved on to NEVs.

Asked, why Chinese authorities did not provide more generous subsidies to CHEVs, a government official explained that since not a single domestic automobile manufacturer was producing them, any new or increased subsidies would be pocketed by Honda and Toyota. As long as Chinese manufacturers did not offer own CHEVs, it was inconceivable for the government to boost support for this technology.

b. Local standardization

The SC has stipulated to "accelerate the establishment of industrial standards conducive to the development of SEIs and a technical standard system for important products" (SC 2010 at 5.3) in the initiation document for the SEI strategy. The definition of national industrial standards and norms is mainly the domain of the MIIT. However, due to complicated deliberations, the release of national standards for grid charging of NEVs was delayed until December 2011 (Cars21 2012). As EV manufacturers had been operating without generally accepted standards for years, they had come to define their own ones. In the absence of national regulation, city governments acquiesced to local companies and promoted their standards. However, since these were tailored to the specific properties and performance parameters of individual products, the resulting patchwork of different—and oftentimes incompatible—standards compromised the evolution of scale economies in core components preventing high cost levels from dropping (Proff et al. 2012).

c. Local protectionism

Chinese industry experts agreed that local protectionism is the biggest obstacle impeding NEV market development. Frustration about this is best illustrated by an article authored by Chinese premier Wen Jiabao that was published in the Chinese Communist Party's bi-monthly official journal Qiushi (Wen 2011). In the paper, Wen expresses dissatisfaction at the oftentimes uncoordinated and incoherent efforts in developing and commercializing electric drive technology. In this regard, he cited administrative failures, such as the tiao-kuai problem, as blocking development progress. He concludes that the direction of development strategies has not been understood by all stakeholders and demands improvements in the coordination mechanisms.

Indeed, integrating sub-central government organizations in the NEV promotion effort greatly complicated the coordination of measures as local authorities have crafted their own development strategies and support incentives. The resulting patchwork of different guidelines reduced the transparency of policy support measures and—more importantly—contributed to the formation of inter-jurisdictional barriers. Fragmentation of the potentially large national market was exacerbated by cities' efforts to reserve local markets for local companies (Sina 2012). Local governments use a large variety of methods to advance "their" companies at the expense of outside competitors.

1. Special incentives were created for encouraging enterprises and individuals to "buy local". Zhengzhou (2011 at 3.3.5), for example, offers purchasing premiums for locally produced BEVs and PHEVs amounting to RMB 60,000 and RMB 50,000 respectively. Dongguan (2010 at 3.3), Chengdu (2010 at 9) and Zhuhai (2011 at 5.3.8) have drafted similar regulations.
2. Municipal governments typically prefer local companies in public procurement and encourage affiliated businesses to follow suit (Ifeng 2012). The governments of Chengdu (2010 at 8), Zhengzhou (2011 at 3.3.4), Nanjing (2012 at 3.8.28), Zhuhai (2011 at 5.3.8) and Hangzhou (2012 at 6.4) have ordered all government departments and affiliated organizations to give preference to locally produced EVs if the latter can meet national industry standards and have equal performance.
3. City governments look to expand their local automotive bases and increase outside sales. The outline of the 12th Five-Year Plan (FYP) for Jilin Province (2011b at 3.2.) stipulates that a major development and production base should be set up with FAW Group at its centre to manufacture 200,000 EVs annually and capture a 10 % share of the domestic market until 2015. Shanghai (2012 at 3.3.1) even aims to reach a national market share of 20 % over the same timeframe.
4. In order to ensure that local suppliers of parts and components capture a sizable share of the market, numerous provinces and municipalities have introduced local content regulation. While some regions, such as Jiangsu (2012 at 1) and Shanxi (2011 at 5.9.) as well as Nanchang (2010 at 5.2.4.3) and Weifang (2010 at 3.5.2), just urge to increase the localization of key components, other jurisdictions specify concrete targets. Anhui Province (2011 at 3.1.) for instance, calls on NEV manufacturers to direct more than 70 % of their procurement volume to local suppliers. Jilin (2011a at 3.1) and Hubei (2012 at 2.7.2) as well as Anyang (2011 at 2.3) set the target ratio at 50 % while Xi'an (2011 at 6.1) demands that over 60 % of inputs are purchased within the city limits. Yet other jurisdictions have drawn up even more elaborate schemes. Zhengzhou (2011 at 3.2.4) and Chengdu (2010 at 3) for example, encourage local NEV manufacturers to purchase locally produced traction batteries, electric engines, power

controls and other key components. As an incentive, authorities pledge to refund a certain proportion of the total purchasing value from local suppliers.

5. Other means of support for local NEV manufacturers include discounts on locally administered vehicle taxes and fees for buyers of local products, as in the case of Nanchang (2010). Zhengzhou has introduced a practice to reward OEMs with special grants for achieving a certain number of NEV sales (Zhengzhou 2011).

For public officials in sub-central government organizations, promoting the local economy takes top priority as the evaluation of their individual work performance is measured by indicators like GDP growth, tax revenues or job creation. In order to ensure career advancement, officials are forced to devote themselves to attract corporate investments which address these requirements (Zhou 2007). NEV manufacturing is highly welcome, as it is expected to stimulate economic growth and technological upgrading of its host cities in a profound and lasting way. Furthermore, as the automotive value chain is among the longest in the manufacturing sector, NEV operations are envisioned to bring along a large numbers of suppliers and related businesses. In terms of employment effects, complete vehicle manufacturing looks exceptionally promising as well: for every job created in automobile production, 11 more will be created in related industries (Sina 2012).

The example of Chengdu shall serve to illustrate local government behaviour. In 2010, authorities announced that supporting policies would be customized to the specific requirements of leading local manufacturers of NEVs, traction batteries, electric motors or power controls (Chengdu 2010, at 3.3.2). In some cases, such as Shenzhen, preferential policies have accelerated the development of "local champions" whose products enjoy special promotion. Despite their short-term benefits for local economies, it is highly uncertain if local champions can grow into national or even global champions as administrative barriers pay an ambivalent role. While protected local markets allow companies to grow in the beginning, the parcelling of the national market keeps companies from realizing scale economies and acquiring genuine competitive advantage (Ifeng 2012).

Among the cities taking part in the *10 Cities 1000 Vehicle Program*, Hangzhou, Suzhou, Zhengzhou and Beijing stood out for their achievements in introducing EVs into their fleets. Collusion between local administrations and companies and "buy local" regulations played a major role in these cases. While the *Program* dictated that participating municipalities had to base vehicle procurement on a formal bidding process (MOF and MOST 2009 at 3.8), the experts consulted held the opinion that government favouritism had been strong and widespread. While several cities organized public tender proceedings, Zhengzhou, Changchun, Wuhan, Xiamen, Guangzhou, Shanghai and four others simply appointed local OEMs to supply the needed vehicles (Gasgoo 2012).

Hangzhou, praised for its relatively high goal attainment of 47 %, had supported local Zotye Motors to supply vehicles for use as taxis. Zhengzhou, Suzhou and Beijing had procured EVs in significant numbers from local bus manufacturers benefitting Yutong (based in Zhengzhou), Kinglong Higer (based in Suzhou) and Beiqi Foton (based in Beijing) (AutoQQ 2012). With the sole exception of

Chongqing, it is instructive that cities with goal attainments below 10 % (Hohot, Haikou, Nantong, Chengdu, Tangshan, Xiangyang and Jinan) neither host major OEMs nor does the automobile industry play a significant role in their economies (21st Century Business Herald 2012b). According to industry experts, these cities are typically unwilling to experiment with promotion measures and assume a wait-and-see attitude instead. Officials are said to observe the developments in other cities to seek out and imitate the most promising approaches. This strategy of playing safe is attractive to many cities without a stake in the auto industry as it prevents the drain of funds through failed investments.

## 6 Conclusion

This paper has demonstrated that the central government has introduced a leapfrogging strategy intended to carry the domestic automobile industry to the forefront of the technological edge. In a clear departure from previous policy orientation, a clear focus was put on NEVs, defined as electric vehicles with power grid connectivity and high electric power ratios. The review of supportive policies has found slow progress in NEV commercialization and development. As a brief overview of the Chinese EV market has illustrated, progress has lagged far behind expectations. In an attempt to identify the underlying reasons for this development, the paper examined the institutional configuration of technological leapfrogging as a state-led approach and shed light on the interaction between central and local authorities in their respective roles as policy initiators and implementers. This review has revealed significant coordination failures both among central level agencies as well as between central and local governments.

On the foundation of a large body of literature on the contentious relationships between central and regional governments in China, this paper has found many of the caveats identified by previous research to be relevant in the current case of NEV promotion. In line with the suggestion of Sood and Tallis (2005) to establish an early lead in a new industry, the Chinese government had started to promote EV development as early as 1999 and designated NEVs a national priority in 2010. However, disputes between central level agencies pertaining to core issues, such as demarcating the precise object of promotion measures or setting national technical standards, delayed the policymaking process—partly offsetting any potential time advantage. Similar to the process described by Lampton (1992), an agreement was reached in principle to kick-start EV development at a relatively early point in time. However, the process got bogged down by attempts to consolidate diverging interests as contentious details had to be negotiated during the implementation phase.

This disunity has also undermined the bargaining position of the central government vis-à-vis regional governments, an issue both Lieberthal (1992) and Lampton (1992) had cautioned against. As suggested by the literature on central-local relations, the quest for local implementation of central initiatives has been cumbersome. However, in the context of the large-scale demonstration projects reviewed here, this finding is somewhat surprising, as participating cities had chosen to do so voluntarily and with generous monetary assistance from the central government's budget. It has been confirmed that policy preferences and development strategies of the various pilot cities diverged widely, mainly depending on the prior existence of local automobile or parts manufacturing. Cities with a stake in the automobile or related industries proved to be far more likely to accommodate NEV promotion policies as the latter complemented or overlapped the cities' own development objectives. Many municipalities without a stake in the automobile industry arguably failed to see the immediate benefit of vehicle electrification and thus dubbed the commercialization of NEVs incompatible to their own economic development strategies. Cities that embraced the NEV development drive sought to integrate the policy into their pre-existing automobile development plans. Unfortunately, this oftentimes brought about results that ran counter to the spirit of the policy. By assigning the task of EV procurement to local automobile manufacturers and other protectionist measures, a number of overzealous municipalities nurtured local champions inside closed off markets. The collusion between local authorities and businesses has led to the fragmentation of the potentially large Chinese EV market and may delay the formation of national champions.

While coordination failures have contributed to the current predicament of the NEV segment, they should not be held solely responsible for the unfavourable market outcome. Following the argument of Schumpeter (1947) and Brezis et al. (1993), the Chinese automobile industry may indeed have been well placed to become a leading player in the future NEVs industry since it had not yet incurred sunk costs related to the preceding technological paradigm to the same extent as its international global competitors. However, following the argument put forth by Wen (2011), Chinese players have thus far failed to show a serious commitment to the new technological paradigm and stake their claim through putting down sunk costs.

Fu and Zhang (2011) have argued that, as a national development objective, TL should be borne by indigenous innovation and independent research. The recent commercial success of foreign EV manufacturers has illustrated, the notion that the Chinese automobile industry could capture a leading position in this future industry based on indigenous innovation might have been overly optimistic. Similarly, by setting the focus on NEVs, authorities appear to have overestimated the rate of technological progress and consumer acclaim in this particular field. Taking note of both factors, central authorities adjusted their approach and assigned CHEVs the role as dominating electric powertrain technology until 2020. Similarly, they have announced to expand coverage of subsidy schemes to Sino-foreign joint ventures and integrate overseas companies in the indigenous innovation scheme. This move signals official acceptance of the idea that Sino-foreign partnerships will become a dominant theme in the future development of the EV industry. In this sense, China's transition to electric mobility may progress without any meaningful changes to the layout of its automotive industry, which has been organized around Sino-foreign joint ventures throughout its modern history. Several experts have pointed out that this approach may be well suited to introduce advanced technologies and speed up the popularization of EVs in China. However, they cautioned against a possible departure of China—as an independent player—from the global competitive arena for EVs. In this case, the country, represented by its market rather than its companies, would be firmly integrated into the global automobile industry which continues to be conducted by the incumbent global players.

## Appendix 1: Buying Incentives for Electric Vehicles

See Tables 1 and 2.

**Table 1** Subsidies provided by cities participating in the project to promote private NEV sales

| City | Subsidies provided |
|---|---|
| Hefei | PHEVs: RMB 15,000 |
| | BEVs: RMB 10,000 |
| | RMB 3,000 reward for gasoline-powered cars being replaced by NEVs |
| | 10 % of central gov't subsidy for purchases of electric micro-buses and trucks by privately-owned companies |
| | 20 % of central gov't subsidy for purchases of electric shuttle buses by privately-owned companies |
| Hangzhou | Purchases of complete vehicles |
| | PHEVs: RMB 2,000 per kWh (max. RMB 30,000) |
| | BEVs: RMB 3,000 per kWh (max. RMB 60,000), minimum battery capacity: 20 kWh |
| | Charging subsidy: RMB 0.09 per km (max. 3 years or 60,000 km) |
| | Rentals of complete vehicles |
| | Charging subsidy: RMB 0.09 per km (max. 3 years or 60,000 km) |
| | Renting subsidy: 50 % of rental costs refunded until yearend 2012 (max. RMB 1,000 per month) |
| | Rentals of vehicle without battery |
| | Renting subsidy 2011: 50 % of rental costs refunded (max. RMB 1,000 per month) |
| | Renting subsidy 2012: 30 % of rental costs refunded (max. RMB 600 per month) |
| | Rentals of batteries without vehicles |
| | Charging subsidy: RMB 0.50 per km (max. 3 years or 60,000 km) |
| | Organized purchases of complete vehicles for staff members (≥10 units) |
| | RMB 3,000 plus RMB 3,000 extra if gasoline-powered cars being replaced by NEVs |
| | Subsidies for installing charging infrastructure |
| | 20 % of the investments (excl. cost for obtaining land) |
| | Subsidies for recycling of spent batteries |
| | RMB 15 per kWh of batteries recycled (only NEV and battery manufacturers) |

(continued)

**Table 1** (continued)

| City | Subsidies provided |
|---|---|
| Beijing | PHEVs: RMB 3,000 per kWh (max. RMB 50,000) |
| | BEVs: RMB 3,000 per kWh (max. RMB 60,000) |
| Shanghai | PHEVs: RMB 2,000 per kWh (max. RMB 20,000) |
| | BEVs: RMB 2,000 per kWh (max. RMB 40,000) |
| Changchun | PHEVs: 20 % of retail price (max. RMB 40,000) |
| | BEVs: 20 % of retail price (max. RMB 45,000) |
| | First 5,000 units: subsidies at 130 % the regular rate |
| | Second 5,000 units: subsidies at 100 % the regular rate |
| | Third 5,000 units: subsidies at 130 % the regular rate |
| | Installation of charging equipment free of charge |
| | Subsidies for companies |
| | Newly established taxi companies using only NEVs in their fleets can claim RMB 2,000 per unit |
| | Battery rental companies receive subsidies for every battery: RMB 10,000 for BEV-batteries and RMB 5,000 for PHEV batteries |
| Shenzhen | PHEV: RMB 30,000 |
| | BEV: RMB 60,000 |
| | Electricity subsidy: RMB 5,000 |

*Sources* Hefei (2012), Hangzhou (2011), Changchun (2011), NEVFocus (2010)

**Table 2** Overview of purchasing subsidies granted for different vehicles (amounts in RMB)

| | | | Maximum electric power rate[a] | | | |
|---|---|---|---|---|---|---|
| | | | BSG[b] | >10 % ≤ 20 % | >20 % ≤ 30 % | >30 % ≤ 100 % |
| Fuel saving | Hybrids (CHEV & PHEV) | >5 % ≤ 10 % | 4,000 | – | – | – |
| | | >10 % ≤ 20 % | | 28,000 | 32,000 | – |
| | | >20 % ≤ 30 % | – | 32,000 | 36,000 | 42,000 |
| | | >30 % ≤ 40 % | – | – | 42,000 | 45,000 |
| | | >40 % | – | – | – | 50,000 |
| | BEVs | 100 % | – | – | – | 60,000 |
| | FCEVs | 100 % | – | – | – | 250,000 |

*Source* MOF and MOST (2009)

[a]The maximum electric power rate (MEPR) indicates the maximum contribution of the electric powertrain components to vehicle propulsion. CHEVs (which use recuperative braking to recharge relatively small batteries) can only count on electric energy to propel the vehicle for a very short distance. Their MEPR is thus relatively low. BEVs on the other hand rely on electric energy for 100 % of the driving distance. Their MEPR is 100 %

[b]A Belt-Driven Starter Generator turns off the engine once a vehicle come to a complete stop and restarts it once the driver releases the brake. The system can recapture energy through recuperative braking and provide the vehicle with a small thrust upon acceleration. As the battery has lower capacity than those other types of hybrid powertrains, the BSG system is not capable to propel the vehicle by itself

## Appendix 2: Classification of Technology Maturity

See Table 3.

**Table 3** Classification for product and technology maturity

| | Starting stage | Development stage | Maturing stage |
|---|---|---|---|
| The route for realizing technical principles is… | Undergoing pre-study | Basically determined | Clearly determined |
| A set of industrial standards is … | Not available | Incomplete | Complete |
| Requirements for commercialization are | Not met | Partially met | Completely met |
| Manufacturing scale | Production in small batches only | Batch production allowed | Regular production |
| Limitations | Demonstration projects in approved regions, within a prescribed time limit and under predetermined conditions | Distribution and sales in approved regions, within a prescribed time limit and under predetermined conditions | No limitations |
| Monitoring | Real-time monitoring for each vehicle | Real-time monitoring for 20 % of vehicles | No requirements |

Based on MIIT (2009) at 7 and 9

## Appendix 3: Timeline for the Planned Transition to Electric Mobility

### Phase I (2008–2010)

- Launch EV pilot projects in public services of large and medium-sized cities.
- Completion of a large-scale pilot project involving 595 battery electric vehicles during the 2008 Olympic Games in Beijing.
- Initiation of the 10 Cities 1000 Vehicle Programme in 2009 as large-scale pilot promotion project in urban centers across the country.

### Phase II (2010–2015)

- Realize the production of hybrid vehicles on an industrial scale.
- Put a cumulative combined total of 500,000 BEVs and PHEVs on the road until 2015.
- Launch large-scale pilot projects for the commercialization of small-size battery electric vehicles

- Initiate small-scale pilot projects for the evaluation of fuel-cell electric vehicles in public services.
- Conduct research and development activities with regard to fuel-cell electric vehicles as the next generation of pure electric vehicles and establish a research and development platform in this area.
- Provide support to science and development activities in order to facilitate the increase of the proportion of pure electric vehicles to all vehicle sales to level of 1 %.
- Promote comprehensive research and development efforts focusing on lithium ion batteries.
- Intensify technology support for the modularization of vehicle batteries as cornerstone of indigenous innovation efforts.
- Reach a breakthrough in the mass production of vehicle batteries.
- Set up 400,000 charging posts and 2,000 stations for battery charging and replacement in over 20 pilot cities and border areas.

**Phase III (2015–2020)**

- Continue to promote the large scale industrial production of pure electric and commence work on the next generation of pure electric vehicles.
- Increase the combined BEV and PHEV population to 5 million units until 2020.
- Achieve a combined production capacity for BEVs and PHEVs of 2 million units by 2020.
- Launch the production of the next generation of vehicle batteries on an industrial scale.
- Optimize the basic support infrastructure for electric vehicles.
- Enhance vehicle-network integration.
- Continue to provide technology support to promote the popularization of all types of electric vehicles until 2020.

Based on SC, 2012.

## References

21st Century Business Herald (2012a) Lianghua Shanglu Mubiao: Duo Di Zhengfu Qiang Tui Xinnengyuan Qiche. http://www.21cbh.com/HTML/2012-2-8/3NNDE2XzM5OTk3Ng.html. Accessed 03 Oct 2012

21st Century Business Herald (2012b) 'Shi Cheng Qian Liang' Gongcheng Luokong? 25 Ge Shidian Tuiguang Mubiao Jin Wancheng 38 %. http://www.21cbh.com/HTML/2012-1-10/1OMDY5XzM5NDQ1OA.html. Accessed 10 Oct 2012

AutoQQ (2012) Xinnengyuan Shidian Shouguan Diaocha Shi Cheng Qian Liang Shifan Gongcheng. http://auto.qq.com/a/20120731/000262.htm. Accessed 15 Aug 2012

AutoQQ (2010) Chang'an Jiexun HEV Shangshi Yi Nian Ling Xiao Liang Beiqing Xiemu. http://auto.qq.com/a/20101207/000049.htm. Accessed 01 Oct 2012

Babbie ER (2010) The basics of social research. Cengage Learning, New York

Brezis E, Krugman P, Tsiddon D (1993) Leapfrogging in international competition: a theory of cycles in national technological leadership. Am Econ Rev 83(5):1211–1219
Cars21 (2012) China releases four EV charging standards. http://www.cars21.com/news/view/727. Accessed 10 Feb 2012
Chen D, Li-Hua R (2011) Modes of technological leapfrogging: five case studies from China. J Eng Tech Manag 28:93–108
China.org.cn (2012) Xinnengyuan Qiche Yingjie Xin Jiyu Buru ‘Huangjinqi’. http://finance.china.com.cn/industry/ny/20120508/706996.shtml. Accessed 30 Aug 2012
Emmel N (2013) Sampling and choosing cases in qualitative research. Sage Publications Ltd, Los Angeles
EVTimes (2012) 2012 Nian 8 Yue Zhongguo Xinnengyuan Qiche Chanxiao Baogao. http://www.evtimes.cn/html/news/q/z/2012/0921/40228.html. Accessed 03 Oct 2012
Flick U (2014) The Sage handbook of qualitative data analysis. Sage Publications Ltd, Los Angeles
Fourin (2012) Private new-energy vehicles: sales lagged below 1,500 units at end of 2011. Fourin China Automot Intell 38:32–33
Fu X, Zhang J (2011) Technology transfer, indigenous innovation and leapgrogging in green technology: the solar-PV industry in China and India. J Chin Econ Bus Stud 9(4):329–347
Gasgoo (2012) Shi Cheng Qian Liang Diao Cha: Fei Shi Dian Hou Lai Ju Shang. http://auto.gasgoo.com/News/2012/09/010119371937864.shtml. Accessed 30 Aug 2012
Howell J (2006) Reflections of the Chinese state. Dev Change 37(2):273–297
Ifeng (2012) Xinnengyuan Che Shifan Gongcheng Nan Dabiao Zuai Chanye Fazhan. http://auto.ifeng.com/news/domesticindustry/20120113/741651.shtml. Accessed 03 Mar 2012
in der Heiden P, Taube M (2012) Development perspectives of the Chinese electric vehicle market. In: Proff H et al (eds) Schritte in die künftige Mobilität – technische und betriebswirtschaftliche Aspekte. Springer-Gabler Verlag, Wiesbaden (forthcoming)
Lampton D (1992) A plum for a peach: bargaining, interest and bureaucratic politics in China. In: Lieberthal K, Lampton D (eds) Bureaucracy, politics and decision-making in post-mao China. University of California Press, Berkeley
Lieberthal K (1992) Introduction. In: Lieberthal K, Lampton D (eds) Bureaucracy, politics and decision-making in post-mao China. University of California Press, Berkeley
Lieberthal K, Oksenberg M (1988) Policy making in China: leaders, structures, and processes. Princeton University Press, Princeton
Meier N (2009) China: the new developmental state? An empirical analysis of the automotive industry. Internationaler Verlag der Wissenschaften, Frankfurt
Oi J (1998) The evolution of local state corporatism. In: Walder A (ed) Zouping in transition: the process of reform in rural north China. Harvard University Press, Cambridge
Proff H et al (2012) The competitiveness of the European automobile industry in the field of electric vehicles. Report prepared for the European Commission (forthcoming)
Schumpeter J (1947) Capitalism, socialism and democracy. Harper, New York
Sina (2012) Xinnengyuan Qiche ‘Jiedong’ Zhengce Hongli Jijiang Xianxian. http://auto.sina.com.cn/news/2012-09-08/10031029172.shtml. Accessed 10 Oct 2012
Soete L (1985) International diffusion of technology, industrial development and technological leapfrogging. World Dev 13(3):409–422
Sood A, Tellis G (2005) Technological revolution and radical innovation. J Market 69:152–168
Steinmüller W (2001) ICTs and the possibilities for leapfrogging by developing countries. Int Labour Rev 140(2):193–210
Unger J (1987) The struggle to dictate China’s administration: the conflict of branches vs areas vs reform. Aust J Chin Aff 18:15–45
Wen J (2011) Guanyu Keji Gongzuo de Ji Ge Wenti. In: Qiushi, No. 14
World Bank (2011) The China new energy vehicles program: challenges and opportunities. http://siteresources.worldbank.org/EXTNEWSCHINESE/Resources/3196537-1202098669693/EV_Report_en.pdf. Accessed 13 Mar 2012

Xinhua (2012a) 2012 Nian Xinnengyuan Qiche 'Shi Cheng Qian Liang' Shouguan Diaocha. http://big5.xinhuanet.com/gate/big5/www.hn.xinhuanet.com/2012-07/24/c_112515881.htm. Accessed 30 Sept 2012

Xinhua (2012b) Jia Xinguang: Shei Lai Kou Dong Chu Fa Hundong. http://news.xinhuanet.com/auto/2012-07/24/c_123458581.htm. Accessed 30 Aug 2012

Yuan D (2002) Jishu Chuangxin de Tezhi Yu Gongneng. Qiushi 6:50–52

Zheng Y (2007) De facto federalism in china: reforms and dynamics of central-local relations. World Scientific Publishing, Singapore

Zhou L (2007) Governing China's local officials: an analysis of promotion tournament model. Working paper of the Guanghua School of Management and IEPR, Peking University

Anhui Province (2011) Anhuisheng Renmin Zhengfu Guanyu Yinfa Anhuisheng 'Shi Er Wu' Gongye Fazhan Guihua de Tongzhi. WanZheng [2011] No. 79

Anyang (2011) Anyangshi Renmin Zhengfu Guanyu Jiakuai Qiche Chanye Fazhan de Ruogan Yijian. AnZheng [2011] No. 43

Chengdu (2010) Chengdushi Diandong Qiche Chanyehua Xingdong Fang'an (2010–2012 Nian. ChengBanFa [2010] No. 17

Dongguan (2010) Dongguanshi Tuidong Diandong Qiche Guanjian Jishu Ji Chanye Fazhan Shishi Yijian. DongFu [2010] No. 80

Hangzhou (2012) Hangzhoushi Renmin Zhengfu Bangongting Guanyu Gongbu 2012 Nian Hangzhoushi Zhengfu Caigou Mulu Ge Xian'e Biaozhun de Tongzhi. HangZhengBanHan [2012] No. 9

Hubei Province (2012) Hubeisheng Zhanlüexing Xinxing Chanye Fazhan 'Shi Er Wu' Guihua. E'ZhengFa [2012] No. 40

Jiangsu Province (2012) Jiangsusheng Jingji Ge Xinxihua Weiyuanhui Guanyu Yinfa 2012 Niandu Xinnengyuan Qiche Chanye Fazhan Xingdong Jihua de Tongzhi. SuJingXinChanYe [2012] No. 225

Jilin Province (2011a) Jilinsheng Guomin Jingji he Shehui Fazhan Di Shi Er Ge Wu Nian Guihua Gangyao. JiZhengFa [2011] No. 6

Jilin Province (2011b) Jilinsheng Zhanlüexing Xinxing Chanye Peiyu Jihua (2011–2015 Nian). JiZhengFa [2011] No. 22

MOF and MOST (2009) Guanyu Fazhan Jieneng Yu Xinnengyuan Qiche Shifan Tuiguang Shidian Gongzuo De Tongzhi. CaiJian [2009] No. 6

MOF, NDRC and MIIT (2011) Guanyu Tiaozheng Jieneng Qiche Tuiguang Butie Zhengce de Tongzhi. CaiJian [2011] No. 754

MOF et al. (2010) Guan Yu Kai Zhan Si Ren Gou Mai Xin Neng Yuan Qi Che Bu Tie Shi Dian De Tong Zhi. CaiJian [2010] No. 230

MOST (2012) Diandong Qiche Keji Fazhan 'Shi Er Wu' Zhuanxiang Guihua. GuoKeFaJi [2012] No. 195

MOST et al. (1999) Kexuejishubu Deng Shisi Bumen Guanyu Yinfa 'Guanyu Shishi 'Kongqi Jinghua Gongcheng – Qingjie Qiche Xingdong' de Ruogan Yijian' de Tongzhi. GuoKeFaGaoZi [1999] No. 564

Nanchang (2010) Nanchangshi Jieneng Yu Xinnengyuan Qiche Shifan Tuiguang Shidian Shishi Fang'an. HongFuTingZi [2010] No. 74

Nanjing (2012) Nanjingshi Zhengfu Guanyu Yinfa Jinyibu Shenhua Shishi 'Qian Qi Sheng Ji' Xingdong Jihua de Gongzuo Yijian de Tongzhi. NingZhengFa [2012] No. 238

NDRC (2004) Qiche Chanye Fazhan Zhengce

NPC (2011) Zhonghua Renmin Gongheguo Guomin Jingji he Shehui Fazhan Zhengce

SC (2012) Guowuyuan Guanyu Yinfa Jieneng Yu Xinnengyuan Qiche Chanye Fazhan Guihua (2012–2020 Nian) de Tongzhi. GuoFa [2012] No. 22

SC (2011) Guowuyuan Guanyu Yinfa 'Shi Er Wu' Kongzhi Wenshi Qiti Paifang Gongzuo Fang'an de Tongzhi. GuoFa [2011] No. 41

SC (2010) Guowuyuan Guanyu Jiakuai Peiyu He Fazhan Zhanlüexing Xinxing Chanye de Jueding. GuoFa [2010] No. 32

SC (2009) Qiche Chanye Tiaozheng he Zhenxing Guihua

Shanghai (2012) Shanghaishi Qiche Chanye "Shi Er Wu" Fazhan Guihua. HuJingXinZhuang [2012] No. 427
Shanxi Province (2011) Shanxisheng Renmin Zhengfu Guanyu Jiakuai Peiyu He Fazhan Zhanlüexing Xinxing Chanye de Yijian. JinZhengFa [2011] No. 21
Weifang (2010) Weifangshi Renmin Zhengfu Guanyu Jiakuai Xinxing Chanye Fazhan de Shishi Yijian. WeiZhengFa [2010] No. 26
Xi'an (2011) Xi'anshi Renmin Zhengfu Bangongting Guanyu Yinfa Xi'anshi 'Shi Er Wu' Gongye Fazhan Guihua de Tongzhi. ShiZhengBanFa [2011] No. 114
Zhengzhou (2011) Zhengzhoushi Renmin Zhengfu Guanyu Zhichi Xinnengyuan Qiche Chanye Fazhan de Yijian. ZhengZhengWen [2011] No. 205
Zhuhai (2011) Zhuhaishi Renmin Zhengfu Bangongshi Guanyu Yinfa Zhuhaishi Xinnengyuan Qiche Chanye Fazhan Guihua (2011–2020) de Tongzhi. ZhuFuBan [2011] No. 11

## Author Biography

**in der Heiden** is working as research associate at the Chair of East Asian Economic Studies at the University of Duisburg-Essen. Previously, he has studied economics and East Asian studies in Germany and China. His main field of research is Chinese economic development with a focus on industrial policy in the metallurgical and automobile industry.

# Part III
# Management Perspectives

# Organizations as Change Agents Towards New Modes of (Sustainable) Mobility: Insights from Practice Oriented Projects

**Dirk Dalichau and Birgit Blättel-Mink**

**Abstract** Nowadays, for many reasons, the transformation of individual mobility behavior seems crucial. There is a consensus among scientists that in time we will run out of fossil fuel. Secondly, in order to at least slow down climate change, pollution of the environment has to be minimized. In addition to these ecological aspects, city centers and urban areas are running out of space in the face of increasing individualized traffic. On the other hand, individual mobility is a crucial factor to enable social participation and integration. Not being mobile often means living on the fringe of society. From an economic point of view, the mobility sector needs to be competitive on a global level but with lower energy use. From the point of view of the customer finally, mobility must be affordable. This contribution focusses on electric cars, on car sharing, and on intermodal services as possible ways of reducing emissions, decreasing the number of cars and overcoming the problem of insufficient space without at the same time increasing social inequality and threatening economic competitiveness. In an ongoing research project, the role of organizations as change agents towards sustainable mobility is being explored. By offering electric vehicles to their employees or tenants, organizations, as employers or landlords, seem to be important actors regarding a successful implementation and diffusion of new modes of mobility. Using a mixture of empirical methods, the project has traced implementation processes of e-fleets in different kinds of organizations and identified different types of promoters from within the organizations as well as structural and institutional barriers to these drivers of change. The preliminary findings allow deeper understanding of how organizations understand their roles in a transformation process towards new modes of (sustainable) mobility, but they also indicate some of the problems that still exist.

D. Dalichau (✉) · B. Blättel-Mink
Department of Social Sciences, Institute for Sociology, Goethe-University Frankfurt, Theodor-W.-Adorno-Platz 6, 60323 Frankfurt am Main, Germany
e-mail: dalichau@soz.uni-frankfurt.de

B. Blättel-Mink
e-mail: b.blaettel-mink@soz.uni-frankfurt.de



D. Fornahl and M. Hülsmann (eds.), *Markets and Policy Measures in the Evolution of Electric Mobility*, Lecture Notes in Mobility,
DOI 10.1007/978-3-319-24229-3_8

**Keywords** Mobility change · Organizations as change agents · e-Fleets · Sustainability

## 1 Introduction

The international scientific community is not only in agreement that in time we will run out of fossil fuel, but also that peak oil and climate change are man–made phenomena. We live in the era of "anthropocene" (Steffen et al. 2007). In order to at least slow down climate change, pollution of the environment has to be minimized dramatically. Individual mobility behavior is among the greatest drivers of climate change. Furthermore, city centers as well as highly agglomerated regions are running out of space in the face of increasing individualized traffic. Transformation of individual mobility behavior seems crucial. On the other hand, individual mobility is a crucial factor to enable social participation and integration. Not being mobile often means living on the fringe of society. From an economic point of view, the mobility sector needs to be competitive on a global level but with lower energy use. From the point of view of the customer, mobility must be affordable. At the end of the day, in order to transform mobility into (more) sustainable mobility we have to overcome these three challenges: ecological, economic and social.

Electric cars are a new means of transportation, not just in the sense that they are powered by electricity instead of fossil fuels. The diffusion of electric cars could also imply a change of everyday mobility. With an average range of no more than 150 km, users of electric cars might decide on a combination of e-vehicles and public transportation system for longer distances. In today's Germany, many regions—especially the highly agglomerated ones—offer integrative mobility systems with intermodal services that make it quite easy and attractive to choose always the 'best' means of transportation. In the centers of larger German cities, more and more people are using car or bike sharing for the fulfillment of their individual mobility needs. Today, many car sharing providers offer, in addition to conventional fossil fuel cars, electric cars. The vision of 'green mobility', combining electric mobility with intermodal services, seems possible (Brand et al. 2012). But electric cars are still perceived as way too expensive for most of the potential customers. Furthermore, people doubt that electric mobility is the leading technology in the mobility system of the future. While some electric vehicles, like e-bikes, are already accepted and sufficiently field-tested, others, such as the e-car, are still perceived as immature (Blättel-Mink et al. 2013; Dütschke et al. 2012).

In 2008, the German Federal Ministry of Transport and Digital Infrastructure initiated a program on e-mobility, aiming to gain new knowledge about how the market penetration of electric mobility could be fostered and intensified. During the first phase of this program users had the opportunity to experience electric mobility under optimized conditions (especially no costs). One finding of the accompanying social-scientific research has been that these experiences influence users' portfolio of means of transportation. Experimental users of electric cars tend to be users of car sharing, bikes or public transportation systems as well (Blättel-Mink et al. 2013).

But who is likely to push the diffusion of electric cars? There is one field in which the diffusion of electric mobility is prospering. In the case of fleets in organizations, the number of electric cars is rising more rapidly than in the mobility sector in general. Some of these organizations are companies in the 'green sector,' such as renewable energy plants, other companies are trying to increase the sustainable impact of their activities (measured in $CO_2$-equivalents), and still others wish to make their 'green engagement' publicly visible (Fichter 2010). In addition to organizations as employers, housing companies are trying to offer new services to their tenants and so constitute another interesting case. Housing companies are installing car pools in their properties, intending to offer housing and mobility as a package and to reduce the number of cars and therefore also the area of parking space needed. Another objective is to offer mobility to those among their tenants who cannot afford their own car.

In the ongoing second phase of the program, the authors, in charge of accompanying research, are striving to gain greater insight into the role of organizations as change agents towards sustainable mobility. Using a mixture of empirical methods, we have traced the implementation processes of e-fleets in different kinds of organizations and identified different types of promoters from within the organizations as well as structural and institutional barriers to these drivers of change.

In a first step, the theoretical sketch of the research project will be presented in two parts. First, we reflect upon individual mobility behavior and ways to transform it. Second, we introduce organizations as change agents. In a next step the research design is described. We then present the preliminary results of our research, which indicate that in many cases the implementation of electric fleets and the initiative to shift mobility behavior in a more sustainable direction is being fostered by the tremendous engagement of individual promoters within organizations.

## 2 Conceptualizing Change of Mobility Behavior

Mobility has to be seen as far more than just moving oneself (or whatever) from point A to point B. Mobility entails symbolic meanings (Featherstone 2004; Sheller 2004; Urry 2007): for example, and very importantly, the freedom to go wherever you want to by whatever means of transportation you choose. Motorized private transport, at least in Germany, promises great advantages in that context.

> The promise here is for self-steering autonomy and capacity to search out the open road or off-road, encapsulated in vehicles which afford not only speed and mobility, but act as comforting protected and enclosed private spaces, increasingly a platform for communications media, that can be enjoyed alone or in the company of significant others. Not only an attractive marketing image in which cars are positioned traversing the wild parts […] of the planet such as deserts and mountain passes, but something which also speaks to

> powerful cultural dreams of adventure and freedom: the capacity to go anywhere, to move and dwell without asking permission, the self-directed life free from the surveillance of the authorities. (Featherstone 2004, p. 1f)

More pointedly: "The car we might say is a way of life and not just a transport system for getting from one place to another." (Urry 2007, p. 115). And it is even more than that. For a long time cars have been part of a complex system structuring our daily lives, especially in a country like Germany where the automobile industry is one of the strongest worldwide (Legler et al. 2009). The automobile sector gives work to thousands of employees, and people tend to identify strongly with products made by the company that employs them. People working in the car industry drive their "own" car. Scientifically, the social or symbolic meaning of the car is of great interest. The symbolism and the emotional affiliation of cars influence the decision making processes of users.

Furthermore, it was the car that made possible the split between place of work and place of residence. This is historically almost unprecedented.

> Social life has become locked into the modes of mobility that automobility generates and presupposes. Something which as the 20th century unfolded has seen a shift from public transport (railways, buses, trams, ships etc.) to the private car. As time-space structures become refigured there is a concomitant shift in forms of sociality, living together and inhabitation. Automobility makes possible the division of the home from the workplace, of business and industrial districts from homes, of retail outlets from city centers. It encourages and demands an intense flexibility as people seek to juggle and schedule their daily set of work, family and leisure journeys, not around the train timetable, but on the calculation of the vagaries of traffic flows. (Featherstone 2004, p. 2)

In the era of peak oil and climate change, a transformation of mobility routines seems crucial. The environment suffers from our current modes of mobility, and space (especially in city centers) is limited. Besides the pure necessity of certain travel needs, e.g. between places of work and residence, which must be rethought in the medium term by people perhaps living closer to their place of work, we need to ask how to break up day to day routines and how to initiate change toward future, sustainable modes of mobility.

But how conceptualize mobility? On the one hand, we can see mobility as a kind of consumption. Classical consumer studies can be used to gain knowledge about the diffusion of new mobility options and the reasons why people accept or reject them. By conceptualizing mobility as consumption, lifestyle becomes an influencing issue as well, implying different styles of mobility (Götz 1997; Deffner 2009; Götz and Deffner 2010; Lanzendorf 2003; Kramer 2005; Harms et al. 2007; Klöckner 2005; Preisendörfer and Rinn 2003). On the other hand, there is the routine-based part of mobility and its possible break-up. Mobility is highly habitualized especially regarding the individual choice of means of transportation (Harms et al. 2007). Therefore the mobility behavior will not be easily altered or changed. The concept of social practices (Reckwitz 2003; Røpke 2009; Shove 2009; Shove et al. 2007; Warde 2005) offers a theoretical understanding of routines in relation to action theory that can be used in the context of consumption. Routine helps to avoid more or less complicated decisions by just repeating internalized solutions for everyday mobility needs. Routine can be seen as an obstacle to transforming mobility (Canzler and Franke 2000). What are the factors that facilitate change in these routines? There are two main types of facilitating factors. Firstly the means of transportation itself, like the damaged car that cannot be used any more or a public transport fare increase. Secondly, life events like relocation, the birth of a child, or a new place of work might lead to new mobility solutions but may also narrow the range of possibilities, due to the fact that not every new situation might fit with every mode of mobility. But what can be done in cases where situations like this do not arise?

Canzler and Franke (2000; 2002) identify three phases that might initiate change in routinized behavior: the search phase, the decision phase and the consolidation phase. During the search phase, users analyze possible variations of their mobility practices by more or less intensive testing of these alternatives, or at least by reflecting on new mobility options and calculating emerging costs. In addition to rational aspects, the social status of different modes of mobility and the symbolic meanings of vehicles play a role. The search phase is followed by a decision making phase in which a specific mode of mobility emerges. This specific mode will be practiced, embedded cognitively and will reach a mental and emotional plateau. It becomes the (new) mobility routine. In the consolidation phase decision makers no longer look for alternatives. The new routine is established.

Information gaps are crucial, especially during the search phase. This extends the search phase and makes an interruption of the three phases likely. An interruption of the process might also happen during the decision phase, e.g. due to an ongoing process of product improvement which is typical for new technologies (such as electric cars). The level of development does not reach a persisting level that makes it easy to decide, so that it always seems reasonable to wait for the next stage of improvement.

Once the change has been made and in spite of the possible obstacles, the combination of a new technology (e-mobility) with the emergence of new practices (car sharing) can be seen as "social innovation" (Howaldt and Jacobsen 2010; Rammert 2010; Zapf 1989).

> Something new, as something that is at first abnormal in society, becomes an innovation when it is offered as an option by more and more social actors or by a relevant social group as a new normality with different rules, or opposed, or recognized, and when this new normality proves to be forward-looking and a change of direction, proves in practice to have consequences that entail change and are superior to present practices, and in terms of its social effects turns out to solve problems better or to lead to progress in some specific respect, and so can be experienced as valuable. (Rammert 2010, p. 34; translation by Gerard Holden)

Having gained an understanding of how mobility behavior is socially embedded and how it can be changed, in the next step we will focus on the organizations that implement new modes of mobility. What is their role during the implementation process, what obstacles arise, and how can organizations exert further impact in terms of changing mobility practices?

## 3 Organizations as Change Agents

Organizations are often providers of mobility, even if it is not their core business. Some examples from organizations that are part of the 'model region' Rhein-Main may illustrate how organizations and mobility are intertwined.

One organization in our sample is a housing company that offers its tenants housing and mobility by implementing car sharing systems in its properties. Other organizations are employers in different sectors that offer their employees electric mobility as part of their car fleets. All the organizations involved in the research project have electric cars as well as e-bike in their fleets. The organizations bear all the so far identified risks of this new technology, such as investing in the 'wrong' technology or not knowing the residual value of the electric car after usage. The users are employees, tenants and customers.

It is obvious that two new social practices come to the fore here: electric mobility and new mobility services. Electric mobility is a technical innovation using a new type of engine to move cars. Car sharing, the second new practice, is understood as a new mobility service and as a social innovation. Very often the participants in our research project are involved in their first contact with shared vehicles—which may be an electric engine, a combustion engine, or a vehicle with no engine (like the conventional bike).

From our research so far we know that gaining experience is one of the key factors that help to increase acceptance of electric mobility among end users. Fun is one of the feelings most frequently mentioned after the first use of an electric car or bike (Blättel-Mink et al. 2013). One of our basic assumptions is that by making such experiences feasible the organizations contribute to the diffusion of electric mobility. At this point we refer to the idea of the 'change agent' (Rogers 2003). While Rogers defines the change agent as an individual, we suggest that an organization might act as or take the role of change agent as well. But let us conceptualize this idea step by step. Rogers defines a change agent as somebody "who influences clients' innovation decisions in a direction deemed desirable by a change agency. In most cases a change agent seeks to secure the adoption of new ideas, but he or she may also attempt to slow the diffusion process and prevent the adoption of certain innovations" (Rogers 2003, p. 312). If organizations act similarly, couldn't they be seen as change agents as well? However, it is individuals who act within an organization and it is individuals who take decisions. Both Witte (1973) and Hauschildt (1997) identify 'promoters' as actors carrying out organizational roles; these are required if changes are to occur. Promoters usually hold positions in organizations that include the possibility of exerting influence on the decision making process. Hauschildt describes different ways of exerting influence. Having specific knowledge about crucial aspects regarding central topics within the organizations is one way of exerting influence as a promoter ('subject promoter'). Holding a position that implicates authority and sway regarding the decision making process is another ('power promoter'). A third type concerns the use of relevant networks and 'tacit' knowledge through the organization's structure, thus leading to influence ('process promoter').

The concept of promoters provides a useful understanding of how individuals, as members of organizations, influence the diffusion of electric mobility in the context of organizations.

## 4 The Research Design

After the conceptual ideas mentioned above had been developed, it became obvious that the project would need to survey both the project coordinators in the participating organizations and the employees as the users of the e-fleets. The project coordinators were asked about how the process of implementing e-fleets in their companies took place, who were the drivers within (as well as outside) the organization, and what kind of problems arose during the process. From a methodological point of view, the project coordinators are understood as 'experts' in the sense that they are the ones who know best about the topic to be explored (Liebold and Trinczek 2009). The method of choice was qualitative expert interviews (Meuser and Nagel 2009; Przyborski and Wohlrab-Sahr 2009). An interview manuscript was developed asking about the process in each phase of implementation, starting with the first idea about how to bring electric mobility into the organization, followed by the implementation process (formulating a grant application, legal regulations etc.) and ending with an assessment of future challenges.

The users of e-mobility within organizations are the people who know best about the advantages and disadvantages of this technology through daily use. Adding their perspective focusses on the question of how to change routines. Is there any change in behavior, and if so what does it consist of? Under what circumstances would the users be ready to change their mobility routines? 'Focus groups' with employees and tenants were set up (Littig and Wallace 1997). Members of focus groups stimulate each other with their user's experiences, which leads to synergetic quality (Henseling et al. 2006). The participants were asked about their most recent mobility behavior, followed by their perception of electric mobility within their associated organization. They were also invited to reflect about a possible usage of electric vehicles as their private affair, and to formulate needs or necessary improvements regarding the situation at their place of work as well as at their home that could influence their decision to use or buy electric vehicles. Finally, they were asked to give a projection of their personal mobility during the next 5 years and to reflect upon possible mobility changes in society as a whole.

The project coordinators are experts on the implementation process, telling us how the process took place and what changes can be observed—in relation to mobility behavior. The users are experts on the actual change in mobility and user based challenges. There is an additional research perspective focusing on the interrelationship between these two spheres. How does the organization initiate change, how is it adopted by the users, and how can it be optimized by both groups collaborating?

## 5 Fostering e-Mobility Within Organizations

The first focus is on the question of how electric mobility makes an entry into the organization. It is obvious that social networks have a role in each of the participating organizations. These networks are initiated by members of the organization interested in electric mobility. Very often that interest is, at least partially, a personal interest in a new technology. Some of the initiators have an affinity with technical innovations in automotive design, while others are interested in finding solutions that will provide more sustainable mobility. If organization members with such interests are equipped with a certain level of competence and influence and if they are in communication with others having the same or similar interests, there is a basis for a successful implementation of electric mobility in the organization. As stated before, these actors can be seen as promoters in Hauschildt's sense of the term (1997).

As 'power promoters', they use their formal position within the organization to initiate change. As 'subject promoters', their particular knowledge of the issue becomes a driver of change. Secondly, their personal interests drive the innovation until it becomes an objective of the organization as a whole. The process of implementing electric mobility can be described as starting with the appropriate persons in the right place at the right time. Therefore, the vision of a new kind of mobility needs to be embedded within the organization. That is why promoters rely on heterogeneous (internal and external) networks, using their potential as 'process promoters' and bringing together people who share an interest in a new technology as well as in a new practice of mobility.

> And ..... that was very interesting, because on the one hand some of the people involved were business people who saw that it could be an interesting business model, and then on the other hand some of us were nerds and were interested in the technology, this is where 20 years ago Volkswagen and Siemens developed, produced and sold the Golf City Stromer. It was a very interesting mixture; people interested in sources of energy just thought it might be very interesting. Producers of cars or suppliers of components, they saw that it could also be a business model, and the technically-minded people were interested anyway..[…]..And there were a couple of environmentalists too. (E-4-01 Free-01; translation)

Formal or informal networks were established interacting with external partners as well. Quite often the initiators, i.e. promoters, of e-fleets relied upon the expertise of partners who already had some experience in the field, whether know-how about charging infrastructure, about the automotive industry or about urban and project planning.

Reflecting upon the preliminary results, it is obvious that sometimes electric mobility only becomes reality by chance. People committed to the new technology or social practice who are connected with internal and external networks and have specific power might be able to start the process. In a next step, institutionalization and professionalization of the implementation process is needed. Only then will electric mobility fleets not be restricted to organizations where employees have specific (technological, sustainable) interests, but can be implemented in every organization where the project makes sense. Professionalization of the implementation process might be another crucial factor to foster the diffusion of electric mobility. This means that it is important not to leave the process to chance but to strategically integrate e-mobility into the outline of the organization. It also means learning from the experimental 'cocoons'.

## 6 Obstacles to e-Fleets in Organizations

It would be a mistake to believe that organizations implement electric mobility concepts by following a rational strategy. Neither promoters nor users act autonomously or without influence from the outside. The 'experimental cocoons' are at the same time highly porous and connected with the organization's environment. In addition to the more subject-driven factors mentioned above, a wide range of external (mainly institutional) and internal (mainly structural) factors influence the implementation of electric mobility. To begin with, the lack of standards regarding technology must be seen as a crucial obstacle. The lack of standards for charging infrastructure[1] prevented the implementation of fast-chargers in at least one organization in our sample. During the grant application process, one coordinator reported that the application needed to include standards for fast charging. As a consequence they had to make decisions on a very tentative basis. Second, and due to the fact that each organization, even if it gets external funding, has to contribute equity capital to a maximum of 50 %, the organization needs security that the best and most promising technology is available. Both the lack of standards and the need for equity capital are obstacles, and in combination they interact. Technologies that lack standards are less likely to be integrated if the organizations have to absorb the risk of wrong investments due to the absence of standardization. One recommendation for funders of projects like this might be to develop new standards by testing different technologies in practice. But the idea of finding or setting new standards by means of funded projects seems questionable as long as practice partners (organizations) have to bring in their own capital, because objectivity towards the 'right' technology will automatically be diminished.

The availability of vehicles is another crucial factor influencing the project designs. Most of the coordinators interviewed stated that they would have preferred electric cars from German manufacturers. But there were only a few German e-mobiles on the market during the first (crucial) months of the projects.[2] This situation caused a higher ratio of foreign models in most of the participating organizations. Taking into account the fact that with this funding program the German government intends to increase the competitiveness of Germany in this new technology, this finding becomes even more significant.

[1]At the time when the grant applications were drawn up, there were still no standards; they were only transposed during the first project phase.

[2]At the end of 2012–beginning of 2013.

Both influencing factors reported so far concern the exterior structures in the context of the funded e-mobility projects. These findings show how the project design in its basic structure—before any car has been bought—is influenced. In the future, these influences could be avoided or minimized by drawing up funding regulations accordingly.

Other obstacles concerning communication on and information about e-mobility among the people involved have been identified. Even in economic organizations mobility is discussed in a very emotional way that does not always recognize the rational pros and cons of new modes of mobility. Two examples should exemplify the lack of rationality during the process of implementing e-mobility in organizations.

In one of the organizations surveyed, a new kind of car pool has been introduced. Up until that point every department of the organization had its own fleet. Among the members of distinct departments a high certainty existed that a car would always be available. The new concept provided a comprehensive car pool for almost all the departments, and reduced the number of cars. This practice provoked worries among the employees about the availability of cars, which was reported by the project coordinators. One coordinator in charge of the implementation, who was responsible for contacting all the departments involved, said that it was not always easy to create trust in the new option among the users. They were worried about the possibility that no car might be available when it was needed, and about the fact that colleagues from other departments used the same pool without knowing each other. The decision made sense from the point of view of costs, but it was criticized by the employees who used the fleet. One reason was that they had not been involved in the decision making process. Only afterwards, when the employees gained experience with the new mobility system, were they reassured, and some of them even became keen on the idea of a shared car pool. During the focus group carried out with the users of the car pool, they confirmed having been concerned about the availability of cars beforehand. But they also reported positive experiences.

Another example concerns employees using a personal company car. When the company offered them an electric car, they were worried about the limited range even though they never drive more than 60–80 km/day. This lack of information became evident during the field research. This obstacle concerns not only the lack of knowledge about the availability of this information but also the willingness to search for this information. But there are some users who are critical of electric mobility in general, without any specific reasons. The following quotation is from a coordinator of a housing company.

> The second reaction is, after they have looked at it more closely: You only have small cars? No better ones? What, they can only travel 120, 150 kilometers? That's no good to me. That's what they say, and when you explain that unless they are going on holiday or on very rare long trips they are only covering distances that are well within the range of the electric cars, they don't believe it. If you can find the time to explain it all in detail, they believe it but they don't want to understand it. And if you talk to them about the cars, they suddenly get worried about safety: Oh, such a small car, I'm much safer in my 5-series BMW. (Interview E-5-02; translation)

Finally, according to the people surveyed the legal regulations about the fiscal benefits of using company cars privately need to be modified. During the first research phase (Blättel-Mink et al. 2013) the use of electric cars was fiscally disadvantageous, but the regulation has now been changed. Nevertheless, the fiscal regulations still favor heavy cars (e.g. SUVs). The price of fuel does not matter either. Not paying for fuel is a common practice in Germany regarding privately used company cars, and is taken for granted by employees at certain levels of a company's hierarchy. These regulations need to be adjusted, and practices in companies must change if a broader diffusion of electric mobility is to be realized.

## 7 Conclusions

Modes of mobility are highly symbolized and status driven, and this implies that a change of mobility towards less polluting engines and new mobility practices will not happen without difficulty. Next to social aspects, technological immaturity and high prices play a role when private individuals think about buying an e-vehicle. Everyday mobility is also highly routinized and hard to break up. Taking all these points together, in addition to the number of good reasons to change behavior towards e-mobility and car sharing, there are also a number of 'good' reasons that hinder these ecologically desirable and necessary transformations. If somebody else takes the risks, can this foster e-mobility and car sharing? In the research project referred to in this paper, the question has been whether the employer who offers an e-fleet to his employees, or the housing company which does the same for its tenants, could take the role of a 'change agent' towards more sustainable modes of mobility. The preliminary results from interviews with project coordinators and from focus group discussions with employees and tenants give positive hints into this direction. In the organizations surveyed, change agency is taken on by the coordinators who promote the implementation of e-fleets. In addition to power and subject promoters, process promoters can be identified who push the project using internal and external support and knowledge. Very often, promotership is intertwined with individual interests in the new technology or with a subjective insight into the need to change mobility. In order to overcome this kind of 'contingency', institutionalization as well as professionalization is needed. This concerns internal as well as external processes. From the point of view of the users—employees and tenants—the successful implementation of an e-fleet needs a culture of participation. Secondly, the fact that companies usually pamper their fleet users, which is even more the case with employees using a personal company car, might be seen as an obstacle to e-mobility and car sharing because e-mobility might not yet offer all the comfort of the combustion engine. Another obstacle in this context is fiscal regulation.

E-fleets might be the rational solution to the ecological problems of today. From the perspective of users e-mobility is often perceived as comfortable and fun, and the acceleration of the electric cars impresses the users. Nevertheless the diffusion of this twofold innovation—technical as well as social—is still at a very early stage. Organizations might have a still underestimated influence on the successful diffusion of electric mobility, which could be optimized if processes within organizations—between promoters and users—as well as contextual conditions are taken seriously. The efforts of organizations as employers or as renters to foster diffusion by offering e-fleets need to be accompanied by legal, informational and technical adaptation. Actually, much more knowledge about electric mobility and its configurations (also within organizations) is available than the actors initiating e-mobility projects in the organizations surveyed assume. They had similar questions during the process of implementation of electric mobility. The specific knowledge needed to answer these questions exists in many cases, as we know from accompanying other electric mobility projects. The reasons why that specific knowledge does not reach some of the project coordinators in the organizations are not yet clear, and further research on this question is needed. To use (or not to use) existing knowledge for the implementation of e-mobility fleets within an organization is also a question of the quality of professionalism.

But at the end of the day, what is necessary is a change in our minds regarding mobility.

## References

Brand M, Loleit M, Braun S (2012) Szenarios zur Elektromobilität 2025. Fraunhofer-Institut für Arbeitswirtschaft und Organisation IAO, Stuttgart

Blättel-Mink B, Buchsbaum M, Dalichau D, Hattenhauer M, Weber J (2013) Transforming mobility into sustainable e-mobility. The example of Rhein-Main region. In: Hülsmann M, Fornahl D (eds) Evolutionary paths towards the mobility patterns of the future. Springer, Heidelberg, pp 113–130

Canzler W, Franke S (2000) Wissenschaftszentrum Berlin für Sozialforschung gGmbH (Ed.): Autofahren zwischen Alltagsnutzung und Routinebruch: Bericht 1 der choice-Forschung. Berlin (Schriftenreihe / Wissenschaftszentrum Berlin für Sozialforschung, Forschungsschwerpunkt Technik - Arbeit - Umwelt, Abteilung Organisation und Technikgenese 02-104). http://bibliothek.wzb.eu/pdf/2000/ii00-102.pdf. Accessed 21 Nov 2012

Canzler W, Franke S (2002) Wissenschaftszentrum Berlin für Sozialforschung gGmbH (Ed.): Mit cash car zum intermodalen Verkehrsangebot: Bericht 3 der choice-Forschung. Berlin, 2002 (Schriftenreihe / Wissenschaftszentrum Berlin für Sozialforschung, Forschungsschwerpunkt Technik - Arbeit - Umwelt, Abteilung Organisation und Technikgenese 02-104). http://nbn-resolving.de/urn:nbn:de:0168-ssoar-112409. Accessed 31 August 2015

Deffner J (2009) Zu Fuß und mit dem Rad in der Stadt - Mobilitätstypen am Beispiel Berlins. Dortmunder Beiträge zur Raumplanung, Verkehr, vol 7. IRPUD, Dortmund

Dütschke E, Schneider U, Sauer A, Wietschel M, Hoffmann J, Domke S (2012) Roadmap zur Kundenakzeptanz. Zentrale Ergebnisse der sozialwissenschaftlichen Begleitforschung in den Modellregionen. Schriftenreihe des Fraunhofer-Institut für System- und Innovationsforschung (ISI). Herausgegeben vom Bundesministerium für Verkehr, Bau und Stadtentwicklung (BMVBS). Bundesministerium für Verkehr, Bau und Stadtentwicklung (BMVBS), Berlin

Featherstone M (2004) Automobilities: an introduction. In: Featherstone M, Thrift N, Urry J (eds) Automobilities. Theory, culture and society, vol 21, no 4–5. Sage Journals, pp 1–45

Fichter K (2010) Nachhaltigkeit: Motor für schöpferische Zerstörung? In: Howaldt J, Jacobsen H (eds) Soziale Innovation. Auf dem Weg zu einem postindustriellen Innovationsparadigma. VS-Verlag für Sozialwissenschaften, Wiesbaden, pp 181–198

Götz K (1997) Mobilitätsstile. Ein sozial-ökologischer Untersuchungsansatz. Arbeitsbericht zum Subprojekt "Mobilitätsleitbilder und Verkehrsverhalten" im CITY: mobil-Forschungsverbund. Institut für Sozialökologische Forschung ISOE, Frankfurt am Main

Götz K, Deffner J (2010) Die Zukunft der Mobilität in der EU. Herausgegeben von Europäisches Parlament. http://www.europarl.europa.eu/studies. Accessed 10 July 2011

Harms S, Lanzendorf M, Prillwitz J (2007) Nachfrageorientierte Perspektive. Das Verkehrsmittelwahlverhalten. In: Schöllgen O, Canzler W, Knie A (eds) Handbuch Verkehrspolitik. VS Verlag, Wiesbaden, pp 735–758

Hauschildt J (1997) Innovationsmanagement. Vahlen, München

Henseling C, Hahn T, Nolting K (2006) Die Fokusgruppen-Methode als Instrument in der Umwelt- und Nachhaltigkeitsforschung. IZT, Berlin

Howaldt J, Jacobsen H (eds) (2010) Soziale Innovation. Auf dem Weg zu einem postindustriellen Innovationsparadigma. VS Verlag für Sozialwissenschaften, Wiesbaden

Klöckner C (2005) Können wichtige Lebensereignisse die gewohnheitsmäßige Nutzung von Verkehrsmitteln verändern? Umweltpsychologie 9(1):28–45

Kramer C (2005) Zeit für Mobilität. Räumliche Disparitäten der individuellen Zeitverwendung für Mobilität in Deutschland. Franz Steiner Verlag, Stuttgart

Lanzendorf M (2003) Mobility biographies. A new perspective for understanding travel behaviour. IATBR Konferenz Luzern. http://www.ivt.ethz.ch/news/archive/20030810_IATBR/lanzendorf.pdf. Accessed 6 Aug 2011

Legler H, Gehrke B, Krawczyk O, Schasse U, Rammer C, Leheyda N, Sofka W (2009) Die Bedeutung der Automobilindustrie für die deutsche Volkswirtschaft im europäischen Kontext. Endbericht an das Bundesministerium für Wirtschaft und Technologie. ftp://ftp.zew.de/pub/zew-docs/gutachten/AutomobEndBericht_final.pdf. Accessed 7 May 2014

Liebold R, Trinczek R (2009) Experteninterview. In: Strodtholz P, Taffertshofer A, Kühl S (eds) Handbuch Methoden der Organisationsforschung. Quantitative und Qualitative Methoden. VS Verlag für Sozialwissenschaften, Wiesbaden, pp 32–56

Littig B, Wallace C (1997) Möglichkeiten und Grenzen von Fokus-Gruppendiskussionen für die sozialwissenschaftliche Forschung. Institut für Höhere Studien (IHS), Wien

Meuser M, Nagel U (2009) Das Experteninterview - konzeptionelle Grundlagen und methodische Anlage. In: Pickel G, Lauth H-J, Jahn D, Pickel S (eds) Methoden der vergleichenden Politik- und Sozialwissenschaft. Neue Entwicklungen und Anwendungen. VS Verlag für Sozialwissenschaften, Wiesbaden, pp 465–480

Preisendörfer P, Rinn M (2003) Haushalte ohne Auto. Eine empirische Studie zum Sozialprofil zur Lebenslage und zur Mobilität autofreier Haushalte. Leske + Budrich, Opladen

Przyborski A, Wohlrab-Sahr M (2009) Qualitative Sozialforschung. Ein Arbeitsbuch. Oldenbourg, München

Rammert W (2010) Die Innovationen der Gesellschaft. In: Howaldt J, Jacobsen H (eds) Soziale Innovation. Auf dem Weg zu einem postindustriellen Innovationsparadigma. VS Verlag für Sozialwissenschaften, Wiesbaden, pp 21–51

Reckwitz A (2003) Grundelemente einer Theorie sozialer Praktiken. Eine sozialtheoretische Perspektive. Zeitschrift für Soziologie 32(4):282–301

Rogers EM (2003) Diffusion of innovations. Free Press, New York

Røpke I (2009) Theories of practice: new inspiration for ecological economic studies on consumption. Ecol Econ 68(10):2490–2497

Sheller M (2004) Automotive emotions: feeling the car. In: Featherstone M, Thrift N, Urry J (eds) Automobilities. Theory, culture and society, vol 21, no 4–5. Sage Journals, pp 221–242

Shove E (2009) Everyday practice and the production and consumption of time. In: Shove E, Trentmann F, Wilk RR (eds) Time, consumption and everyday life. Practice, materiality and culture. Oxford, New York, pp 17–34

Shove E, Watson M, Hand M, Ingram J (2007) The design of everyday life. Oxford, New York
Steffen W, Crutzen PJ, McNeill JR (2007) The anthropocene: are humans now overwhelming the great forces of nature? In: Ambio, vol 36, no 8. Royal Swedish Academy of Science, pp 614–621
Urry J (2007) Mobilities. Polity, Cambridge
Warde A (2005) Consumption and theories of practice. J Consum Cult 5(2):131–153
Witte E (1973) Organisation für Innovationsentscheidungen: das Promotoren-Modell. Schwartz, Göttingen
Zapf W (1989) Über soziale Innovationen. Soziale Welt 40(1–2):170–183

## Author Biographies

**Dirk Dalichau** studied sociology and is scientific assistant at the Department of Social Sciences at Goethe-University Frankfurt/Main, Germany. He recently finished his Ph.D. thesis about standardized consumption conditions and their influence on consumption practices. His main research interests: (sustainable) consumption; transdisciplinary research on sustainability; consumption and experience; prosumption and user integration; electromobility.

**Birgit Blättel-Mink** is Professor for Sociology with main focus on industry and organisation at the Department of Social Sciences at Goethe-University in Frankfurt/Main, Germany. She holds a Ph. D. in Sociology from University of Heidelberg. Her main research fields: (open and social as well as economic) innovation, sustainable consumption. Gender (in)equality in higher education and transdisciplinarity.

# Competition, Cooperation, or Both? Designing Coopetitive Structures in the German Automotive Supply Industry

**Lukas Schilling, Thomas M. Fojcik and Konrad Wirmer**

**Abstract** In the past, cooperation in the automotive industry was observed particularly in the vertical dimension of the value chain. Presently, an increasingly strong trend towards cooperation on the horizontal level and thus between direct competitors is emerging. The large number of examples in the automotive supply industry shows that automotive suppliers are trying to combine resources and competencies with direct competitors to gain competitive advantages. In this context coopetition occurs, i.e. the simultaneity of cooperation and competition between at least two companies. One of the main conflicts of coopetition is the emergence of opportunism among the participating parties, which represents one of the greatest threats to successful partnerships in the automotive industry. Conversely, opportunism is required for competition and thereby for the generation of competitive advantages for at least one side of the cooperation. Thus, the consideration of opportunism plays an important role in coopetitive situations, because a distinct (weak) opportunism exerts a positive (negative) influence on competitive behavior, affecting at the same time negatively (positively) on cooperative activities. In the course of coopetition the challenge to ensure a balance of opportunistic behaviors between cooperative and competitive actions becomes crucial. Against this background, the present article analyses the process and the effectiveness of various organizational measures to balance opportunistic behavior in coopetitive situations in the context of the automotive supply industry. In a first step, various organizational measures of the transaction costs, the principal-agent and social exchange theory were extracted. In a second step, ten German automotive suppliers have been examined during coopetitive situations by means of a case study analysis, in order to test the process, the efficiency and the effectiveness of the theoretically derived measures. Our analysis shows that widespread measures in the

L. Schilling · T.M. Fojcik (✉) · K. Wirmer
University of Duisburg-Essen, Lotharstraße 1, 47057 Duisburg, Germany
e-mail: thomas.fojcik@uni-due.de

L. Schilling
e-mail: lukas.schilling@stud.uni-due.de

K. Wirmer
e-mail: konrad.wirmer@stud.uni-due.de



D. Fornahl and M. Hülsmann (eds.), *Markets and Policy Measures in the Evolution of Electric Mobility*, Lecture Notes in Mobility,
DOI 10.1007/978-3-319-24229-3_9

literature, such as formal contracts or contract-based sanctions, are inefficient and ineffective to balance opportunistic behavior. In contrast, measures which ensure a context of cooperative and competitive activities e.g. social norms and values as well as specific investment and mutual equity investments, are more promising to put opportunism in coopetitive situations into balance.

**Keywords** Automotive supply industry · Competition · Cooperation · Coopetition · Social exchange theory · Opportunism · Principal-agent theory · Transaction costs theory

## 1 Introduction

In the automotive industry a trend to a stronger networking is becoming apparent. In order to gain competitive advantages, companies combine their potentials and resources through cooperation. In the past, cooperation were mostly limited on the vertical dimension. Now companies are cooperating more and more with their direct competitors and thus on the horizontal level of the value chain. Manifold examples show that cooperative and competitive activities are thereby compatible. Coopetition as simultaneously competing and cooperating allows to combine advantages of both types of behavior (Brandenburger and Nalebuff 1996; Wilhelm 2011; Walley 2007; Bengtsson and Kock 2000). Especially small and medium-sized enterprises are able to expand limited resources and competencies in this way. Coopetition favors technological developments, which the automobile companies couldn't accomplish alone. Additionally, it enables that complementary products could be connected for system solutions, markets and products could be expanded, fixed costs could be divided, capacity constraints could be avoided or economies of scale could be generated. Especially in the transition to electric mobility, coopetition has a weightily meaning, because automobile companies can combine resources and competencies with direct competitors through coopetitive behavior in order to ensure adequate economies of scale and scope, to reduce R&D costs and to share supply-side risks in the transition to electric mobility. In this regard coopetition may indirectly lead to a reduction of the sales prices of electric vehicles—which primarily prevent a wide market penetration—and thus contribute to an accelerated diffusion of electric vehicles. Although coopetitive behavior for automotive companies in the transition to electric mobility holds numerous possibilities and opportunities, coopetition provides a very complex form of relationship between direct competitors (Bengtsson and Kock 2000; Dowling et al. 1996). In the context of coopetition the opposite management logics and behaviours lead inside as well as between the participating companies to tensions and role conflicts (Bengtsson and Kock 2000). In addition, goals and interests of the involved companies are variable and because of the competitive character rarely congruent (Nault and Tyagi 2001). On the one hand competitive activities lead to opportunism, which is regarded as the origin of failures of cooperation (Das and Rahmann 2010). On the other hand, cooperative behavior minimizes opportunism, but inhibits advantages generated through competitive activities. Thus the coordination of these opposite logics and behaviors creates a complex framework for companies (Bengtsson and Kock 2000). For a successfully realization of coopetition, companies must therefore identify appropriate conditions and measures, which ensure cooperative and competitive actions simultaneously (Das and Teng 1998). Insofar, a balance of opportunistic attitudes between competitive and cooperative activities should be found. Although this balance of opportunistic attitudes has a high importance for coopetition, earlier examinations were primarily focused on the definition and the terminology of coopetition (Mariani 2007). Operational solutions were indeed required in praxis, but hardly taken up (Bonel and Rocco 2007).

Against this background, we investigated the process as well as the efficiency and effectiveness of different organizational measures to balance opportunistic behaviour in coopetitive settings. In a first step, we extracted different organizational measures from the transaction cost, the principal-agent and the social exchange theory. In a second step, we analyzed German automotive suppliers in the context of coopetitive situations in order to determine the process as well as the efficiency and effectiveness of the extracted measures to balance opportunistic behavior. Our analysis showed that restrictive measures, like for example formal contracts, are inefficient and ineffective for the balancing of opportunistic behavior. The analysis and the results of this study are presented in the following article.

## 2 Design Options of Coopetition

In recent years coopetition as a synonym for the simultaneous realization of cooperative and coopetitive activities between at least two companies, has gained importance in practice and science (Mariani 2007; Walley 2007). Although the simultaneity of both behavior patterns is opposite and contradictory, initial studies indicate that companies can achieve benefits from competition and cooperation through coopetitive structures (Holmlund and Kock 1998). Previous research indicates that a temporal separation, structural separation or a contextual integration of cooperating and competing activities suits well to opposite logics and behaviors (Poole and Van de Ven 1989 and Fig. 1).

Through temporal separation companies can isolate cooperative and competitive activities in a way that only cooperative or only competitive activities are performed at one point of time. In the temporal separation, contradictory logics and behaviors never occur simultaneously—with the exception of the transition period from cooperative to competitive activities, and vice versa. In the context of structural separation, on the other hand, cooperative and competitive activities are separated organizationally. Therefore, a business division can perform purely cooperative actions, while other divisions act competitively. In the context of structural separation, cooperative and competitive activities are indeed implemented

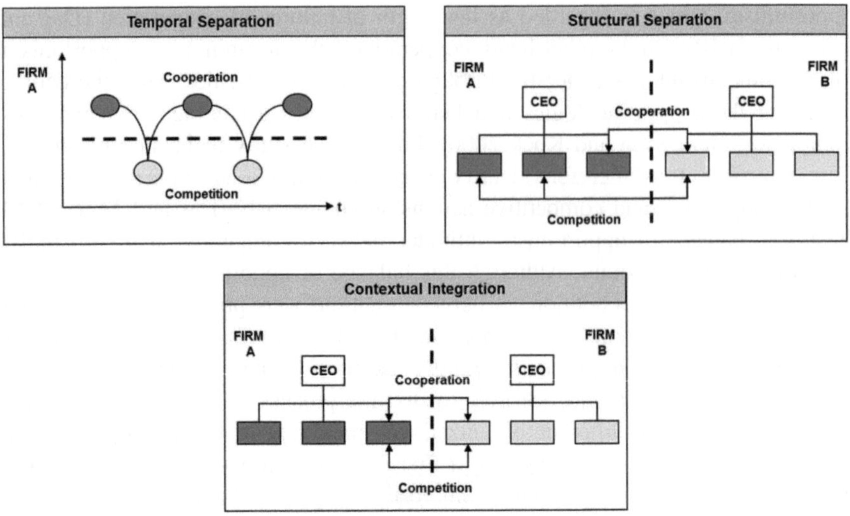


**Fig. 1** Design options of coopetition. *Source* Own illustration

simultaneously on the corporate level, but on the business unit level they are structurally separated (Kotzab and Teller 2003; Dowling et al. 1996). Another way to coordinate conflicting logics and behaviors constitutes the contextual integration of cooperative and competitive activities. In this approach, cooperative and competitive behavior is not separated temporally or structurally, but rather implemented simultaneously on an organization level. For this purpose, a context is necessary, which promotes and ensures on the one hand cooperative and on the other hand competing activities (Fojcik and Schilling 2013). Consequently, the contextual integration reflects a balance between cooperation and competition, whereby this coopetitive structure can also be understood as a pure form of coopetition.

Although all three design options should counteract potential tensions and conflicts due to unclear role allocations under coopetitive situations, the structural separation of competitive and cooperative activities only leads to a shift of contradictory logics and behaviors to a higher management level. From coopetition resulting conflicts and tensions are therefore not eliminated by a structural separation, but rather transferred to other organizational hierarchies. In the temporal separation, competing and cooperative activities are separated in time, whereby tensions and conflicts can be counteracted. However, possible adjustment costs that arise in a constant change between cooperation and competition may reduce the benefits of coopetition. Comparing with the previously outlined design options, the contextual integration rather shows the advantage that competing and cooperative activities can also be implemented simultaneously without adjustment costs or a shift of conflicting behaviors on different hierarchy levels. Although the contextual integration of cooperation and competition keeps positive assets, opportunistic behaviors must also be balanced in the context of this design option. Measures which on the one hand enable, but on the other hand also inhibit opportunistic behavior must be identified to ensure coopetition.

# 3 Opportunism and Measures Towards Opportunistic Behavior

The appearance of opportunism and measures against opportunistic behavior are in essence discussed in the transaction cost theory, principal-agent theory and social exchange theory.

## *3.1 Opportunism in the Transaction Cost Theory*

The transaction cost theory assumes a natural representation of the self-interest of all economic agents. This can be subdivided, depending on the characteristics, in three areas: obedience, simple pursuing of the self-interest and opportunism (Williamson 1990). Williamson defines opportunistic behavior as the "pursuit of the self-interest with the aid of artfulness". He counts "lies, stealing and betraying" plus "deception" to it. Beyond opportunism there exists a complete obedience, which does not correspond to the pursuit of the own interest, but also obtains the plain pursuit of the own interest without the aid of artfulness. In the consideration of the relation between organizations, hart bargaining and high frequency disagreements as well as conflict behavior are not seen as opportunism (John 1984; Provan and Skinner 1989). In the same way, advantages through resources and know-how are not justified with opportunism. In fact, advantages, which are deduced from opportunism, must be induced from a manipulation of information or a faulty presentation of intentions (Provan and Skinner 1989). Despite of the assumption that every individual tries to maximize the own benefit and that companies act in situations of potential advantages always opportunistic, opportunism must not be permanent nor always existent. The transaction cost theory rather supposed that every individual and thus also organizations tend to opportunism in specific situations (Williamson 1990). This constitutes that companies exist, which are not acting opportunistic, but they are, however, difficult and cost intensive to identify (Deeds and Hill 1999).

### 3.1.1 Measures Against Opportunism in the Transaction Cost Theory

The primary mechanisms of the transaction cost theory to contain opportunistic behavior are contractual and structural disincentives. These mainly serve to reduce unilateral advantages that would be generated through opportunistic behavior (Deeds and Hill 1999). The basis of the structure of a business relationship is always a formal contract. Formal contracts are used in the transaction cost theory to coordinate the structure of pay-outs and for the penalty of opportunistic behavior with the goal of bilateral (both-sided) advantages. Contracts are composed of promises or commitments for specific actions and should limit opportunistic behavior through punishments and sanctions (Blumberg 2001). According to the transaction cost theory contracts act as a coverage against risks. It is the main task of the management to cover business relations against these risks and to ensure that the results of the cooperation harmonize with the desired aims. The higher the risks, the more extensive the contract must be designed. This is in turn associated with higher costs (Poppo and Zenger 2002). In contrast to that, other authors point out that the level of detailing in contracts has not an influence on opportunistic behavior (Achrol and Grundlach 1999). To eliminate opportunistic behavior contracts should be completely, clearly and easy to implement. Beyond, the results and outputs of opportunistic actions can be influenced ex-post by certain safeguards. To influence opportunism already ex-ante, additional precautions are required. By the transaction cost theory the safeguards are obtained with a suitable control structure, which is according to Williamson (1981) subdivided into a control, an observation and a measurement function. Observation and control in combination with contractually organized sanctions reduce opportunistic trends (John 1984). At the same time a simple structure should be used in business relationships, as high complexity would lead to unnecessary costs. Otherwise, complex relations require an extensive control structure (Williamson 1979). Basis of the control structure are appropriate coordination and control mechanisms.

Control mechanisms can contain opportunistic tendencies (Wathne and Heide 2000) as they raise the chance to detect opportunistic behavior at an early stage. They help to identify opportunism, which arises from information asymmetries and remains often undetected without control. The coordination of activities, resources and information flows between the partners reduces opportunism by avoiding alignment problems as well as faulty or absent information flows and improves the goal convergence (Crosno and Dahlstrom 2008). The risk of opportunism can be reduced by defining clear processes, which are easier to control. Control and coordination are initiated through formalization and centralization (John 1984). Centralization means in this context the concentration of the decision-making power, for example in the form of a centrally organized decision committee. Contrariwise, formalization is the use of specific procedures and methods in order to control a relationship. Beyond control and coordination mechanisms, specific investments and own capital assets are also seen as an appropriate instrument to contain opportunistic behavior. Specific investments are expenditures, which have no value beyond the business relation. Are specific investments made, a party cannot finish the relationship without accepting economic losses. Specific investments can be used by the other party as a pressuring medium so that opportunism can be prevented in the investing parties (Crosno and Dahlstrom 2008). Very high specific investments can thus also lead to dependency, which in turn prevents opportunism (cf. Hawkins et al. 2008). However, non-investing parties can exploit this situation and expropriate the investing parties. This in turn would be an opportunistic action (Rokkan et al. 2003). Thus, specific investments exert indirect influence on opportunism and can therefore limit opportunistic behavior of the investing party (Mysen et al. 2011). But at the same time specific investments give the non-investing party an incentive for opportunistic behavior (Crosno and Dahlstrom 2008). Other mechanisms of the transaction cost approach are equity investments. The transaction cost theory sees in the purchase of shares the possibility to obligate companies together, whereby opportunism can be reduced. Because equity investments are not recoverable in the short-run, opportunistic behavior would lead to losses under these circumstances (Deeds and Hill 1999).

## 3.2 *Opportunism in the Principal-Agent Theory*

In the principal-agent theory, opportunism is not an explicit mentioned behavior. Nevertheless, the principal-agent theory considers opportunistic behaviors. The principal-agent theory supposes that opportunistic behavior arises when opportunities are offered. A distinction is made between two characteristics of opportunism. The first applies the moral risk, which is based on efforts and achievements of the agent. It is assumed that the agent does not pursue the agreed effort, but goes the less costly way and thus minimizes its use. The second characteristic describes that an agent (often) presents himself better then he really is. The focus of the principal-agent theory is thereby the design of contracts. These should control, as effectively as possible, the relationship between parties, by taking account environmental uncertainty and the impossibility to supervise completely the behavior of the respective opposite party. In this context, the effectiveness of contracts and measures is always considered from the principal perspective (Bergen et al. 1992).

### 3.2.1 Measures Against Opportunism in the Principal-Agent Theory

The principal-agent theory presents various ways how to deal with opportunism within contracts. Considering the objective divergence between agent and the principal as a cause of opportunistic behavior, the principal expects—to achieve best possible results—actions that are connected with great efforts and a large time commitment of the agent (Bergen et al. 1992). In order to prevent that the agent eludes these in an opportunistic way, incentives can be integrated in the contract, by paying the agent on the basis of its performed results. This leads to the fact that the agent has an incentive to behave optimally or to pursue pre-defined goals (Eisenhardt 1989). Another possibility is that the principal procures more information about the behavior of an agent by using, for example, appropriate supervisory measures. Based on this information, the principal can set up a contract, which fits to the behavior of the agent. According to the principal-agent theory, both measures are able to prevent a goal divergence. The monitoring of an agent is also used for the solution of information asymmetries between the agent and the principal. In this case the agent is evaluated by observation and control. Another option are contracts, which are based on the results of the agent's work. The actions of the agent are valued with the help of the achieved results, e.g. the production volume. On this basis result-orientated contracts can be implemented, so that the agent is paid for the effectiveness of his actions (Bergen et al. 1992). Result-orientated contracts serve that the agent acts according to the interests of the principal. Due to the strong focus on final results, the information asymmetry thus gets a lower priority.

## 3.3 *Opportunism in the Social Exchange Theory*

For the examination of opportunism, representatives of the social exchange theory mostly go back to the contribution of the transaction cost theory and respond critically to their assumptions (Grundlach et al. 1995; Heide and John 1992). Contrary to the assumption of the transaction cost theory that opportunism is pervasive, the social exchange theory assumed that in relationships, which are embedded in a social structure, opportunism rather constitutes the exception (Granovetter 1985). In the social exchange theory long-term and successful relationships are primarily based on trust, whereby opportunism represents a significant risk for the creation of trust and therefore increases the potential for conflicts (Luo 2007). In order to maintain long-term and successful relationships, according to the social exchange theory, a structure must be created which prevents uncertainty and opportunism (Grundlach et al. 1995).

### 3.3.1 Measures Against Opportunism in the Social Exchange Theory

In the social exchange theory measures besides trust such as target equality, power symmetry, reciprocity, solidarity, an interest in the common benefit, role integrity, flexibility and a tendency to constructive conflict resolution are considered as opportunism damping parameters. In order to give rise to these factors and prevent or reduce opportunism by the embedment of the business relationships in a social structure, behavior standards and a constant communication should be integrated in the relationship. These are the essential building blocks to limit opportunism (Crosno and Dahlstrom 2008). Communication should prohibit information asymmetries and also should enable goal-arrangements. Furthermore, relational contracts and norms for the control of opportunistic behavior are recommended. Relational norms are hereby fixed values and behavior agreements that are necessary to build up trust. Typically, the norms describe the above mentioned behavior agreements like loyalty, solidarity, interest in the common benefits (reciprocity), willingness to take obligations (role integrity), adaptation (flexibility) and the tendency for constructive conflict solutions (Achrol and Grundlach 1999; Hawkins et al. 2008). By setting exact boundaries for a behavior, standards are a protection against uncooperative behavior. They strengthen the search for common benefits, wherewith opportunism can be reduced (Crosno and Dahlstrom 2008). Furthermore, a relationship, which is characterized by relational norms, is less prone to asymmetries of motivation and thus for opportunistic behavior (Achrol and Grundlach 1999). Thus, norms and values represent an optimal way to control a business relationship (Deeds and Hill 1999).

Relational contracts go back to Macneil (1980). He understands contracts as [...] no more or no less than the relations among parties to the process of projecting exchange into the future (Macneil 1980). The basis of his argumentation is that every relational exchange is based on a social component, especially trust. Relational contracts are formally incomplete and rather serve as a framework for the business relationship (Wathne and Heide 2000) in order to grant the contracting parties, so that they can flexibly adapt the contract to new initial conditions (Jeffries and Reed 2000). In this context the contract is not composed of exact rules and instructions, but rather of the formulation of goals and general applicable arrangements or measures that will be implemented in the event of unforeseen events (Milgrom and Roberts 1992). The strength of relational contracts justify Carson et al. with better efficiency in the restriction of opportunism and higher flexibility by lower cost at the same time (Carson et al. 2006). Formal agreements must be formulated ex-ante, so that they are ex-post verifiable by a third party. Relational contracts on the other hand can build on results that are only verifiable ex-post by the contracting parties. A relational contract must not be accessible for a third party and can contain specific information, which are only comprehensible for the involved parties. In addition, they can be adapted in case of environmental changes or new goals. Due to the unenforceability by third parties, relational contracts must be self-enforcing. For the enforceability the future value of the cooperation must be high so that a breach of contract is not pursued by the partners (Baker et al. 1998).

Although all above-mentioned theoretical approaches describe helpful measures (see Fig. 2), they are limited to cooperation and therefore to settings which largely reduce opportunism. Especially measures from the transaction cost theory and principal-agent theory intend to minimize opportunistic tendencies through supervision and control. A consideration of opportunism in a positive way—meaning that opportunism can also result in competitive advantages—has been largely neglected. In coopetitive situations a complete elimination of opportunism is counterproductive. With the perspective to reach a balance of opportunistic behavior in coopetition we analyzed coopetitive situations in the German automotive supply industry in order to identify measures, which promote as well as arrest opportunistic behavior and are therefore able to balance behavior.

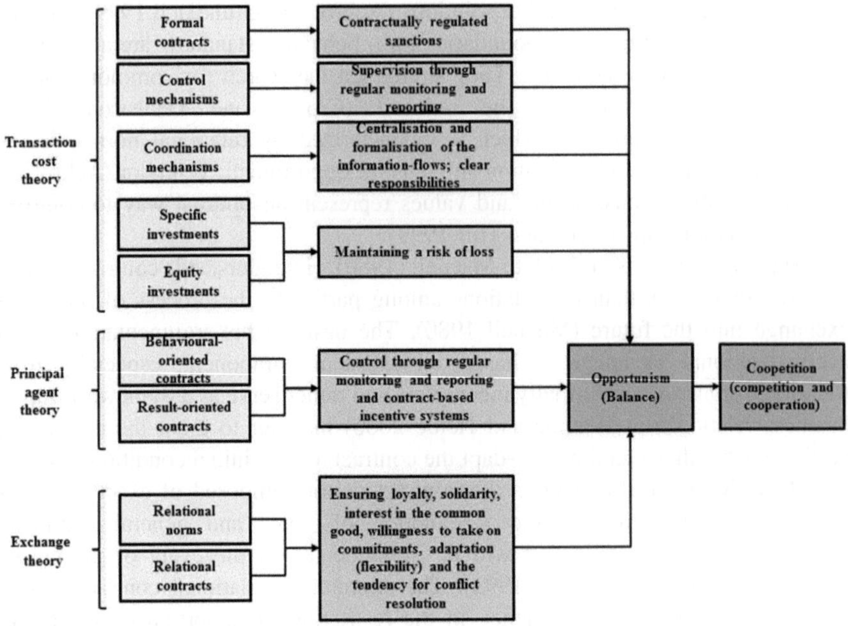


**Fig. 2** Theoretically derived measures against opportunism. *Source* Own illustration

## 4 Analysis: Data Collection and Methodological Structure

The investigation of coopetitive situations in the German automotive supply industry took place between January–March 2013. The examination was restricted to German automotive suppliers in order to avoid possible industry- and country-specific influences. Given that the research field of coopetition has only been studied insufficiently until now and that there is also rarely knowledge about measures to balance opportunism, we chose an inductive approach in terms of a case study analysis. In a first step, 40 German automotive suppliers were selected randomly from the basic population, consisting of firms which showed, with the help of publicly available information, first signs of coopetitive behavior including a contextual integration of competition and cooperation. All 40 automotive suppliers were then invited to telephone interviews. 16 automotive suppliers agreed to participate in an interview. After the interviews, six automotive companies were excluded from further investigation, as they were not in a coopetitive setting according to the research definition. The telephone interviews dealt with coopetition in general and spanned semi-standardized question to problems, factors and measures which cover opportunistic behavior in coopetitve situations. With the help of the interviews the application of the theoretically-derived measures against opportunistic behavior could be analyzed. In addition new measures to balance opportunistic behavior in coopetitive situations could be generated by the semi-standardized questions. To analyze the data base, we used the case study method according to Eisenhardt (Eisenhardt and Graebner 2007).

In order to meet the requirements of the case study method according to Eisenhardt, the final examination included 10 cases, which were subdivided into two groups with five automotive suppliers. The first group incorporated automotive suppliers as cases, where a coopetitive situation was present and had been evaluated by the company representatives as successful (positive cases). However, in the second group were only automotive suppliers as cases, where the coopetitive situation has already been resolved or was indeed still prevailed, but was marked as unsuccessful (negative cases). By contrasting and comparing the two groups, measures and processes to balance opportunistic behavior could be identified, which resulted in a successful design of coopetition. To better understand the mechanisms and the effects, internal and external sources of information had been collected in addition to the interviews. Then all collected information were triangulated in course of the data analysis and merged in a process model, which indicates how the investigated companies implemented measures to integrate cooperative and competitive activities.

## 5 Results of the Case Study Analysis

The results of the case study analysis indicate that automotive suppliers mainly use formal contracts in coopetitive situations in order to reduce opportunistic behavior. This became apparent in the positive, but also in the negative cases. Although a causal conclusion about the effectiveness of formal contracts on the balancing of opportunistic behavior cannot be made, a closer examination of the negative cases shows that formal agreements were rarely used to balance, but primarily used to reduce opportunism. In this context, formal agreements were seen as a last resort in order to limit competitive activities. In most cases, formal agreements led to the prohibition of cooperative and competitive clearances. Finally the coopetitive situations of the participating parties were abandoned because of barely visible benefits from cooperation or competition. Furthermore, the investigation shows that strict and rigid control mechanisms are hardly suitable to balance opportunism. In all negatives cases, the implementation of comprehensive monitoring and supervision measures resulted in a discontinuation of the coopetitive cooperation between automotive suppliers, because the monitoring and supervision measures reduce the mutual trust of the parties in the cooperation. On the one hand, the control mechanisms led to an increase of cooperative activities between the involved companies. But on the other hand they also minimized or prevented almost completely the generation of advantages from competing activities. More promising was a regular reporting system, which was designed so flexible that critical information against the participating partners could be held back. Insofar, the automotive suppliers tried not to eliminate potential information asymmetries, but rather to ensure the presence of an opportunistic attitude so that competing activities were still possible.

Contrasting the positive and negative cases, it became apparent that specific investment and mutual equity investments may indeed assure coopetitive situations. In contrast to the positive cases, no or only one-sided investments and equity investments have been made in the negative cases. That's why automotive suppliers were neither able to ensure a commitment to mutual consumption of cooperative advantages, nor to reduce reservations and barriers against the sharing of resources, competencies and competitive advantages. Without specific investments, automotive suppliers could rarely balance opportunistic behavior because the complete consumption of opportunistic benefits was not accompanied simultaneously with the risk of loss of specific investments and equity participations. In most of the cases coopetition was therefore discontinued in favor of competing activities. In contrast, the positive cases show that centralization and mutual investments favored clear responsibilities and a strict formalization of the information flows between the participating automotive suppliers. These measures in particular promoted the awareness of a joint objective, facilitated the communication between the parties and signaled the importance of coopetitive behavior to the directly affected business area through coordination at the highest hierarchical level. Consequently, opportunistic behavior was not completely prevented by these measures, but limited in its scope on the business area and employee level. For the support of these coordination mechanisms automotive suppliers introduced further accompanying measures, which successfully transposed coopetition. Especially behavioral and results-oriented incentives for directly involved employees should reduce opportunism at the individual level. Investigating the negative cases, it became interestingly apparent that the previously described coordination mechanisms and accompanying measures were in part or completely missing in this group of automotive suppliers. In contrast to the positive cases of the study, these automotive suppliers also neither inserted relational norms nor relational contracts. Important relationship factors for coopetitive behavior such as loyalty, solidarity, interest in the common good, willingness to take over commitments or the affection for constructive conflict solutions that are positively influenced by relational norms and contracts could not be guaranteed in that manner. Consequently, these automotive suppliers were not able to ensure a social context between the involved companies which could balance opportunistic behavior and therefore enables coopetition.

Besides the measures for balancing opportunism and thus for ensuring coopetitive behavior, the positive cases showed similarities in the temporal implementation of the respective individual measures. In summary, three different phases (meeting phase, transformation phase and aggregation phase) were identified, how automotive suppliers control opportunistic behavior in coopetitive situations (see Fig. 3).

In the meeting phase economic respectively legal independent automotive suppliers have the possibility to become acquainted with each other. In this phase they can communicate about goals and interests and can identify potential conflicts as well as problems. Furthermore, automotive suppliers mostly collaborate in uncritical areas, without taking too much risk. Thus, the meeting phase has a preventive and reactive nature towards opportunism. Opportunistic behavior will

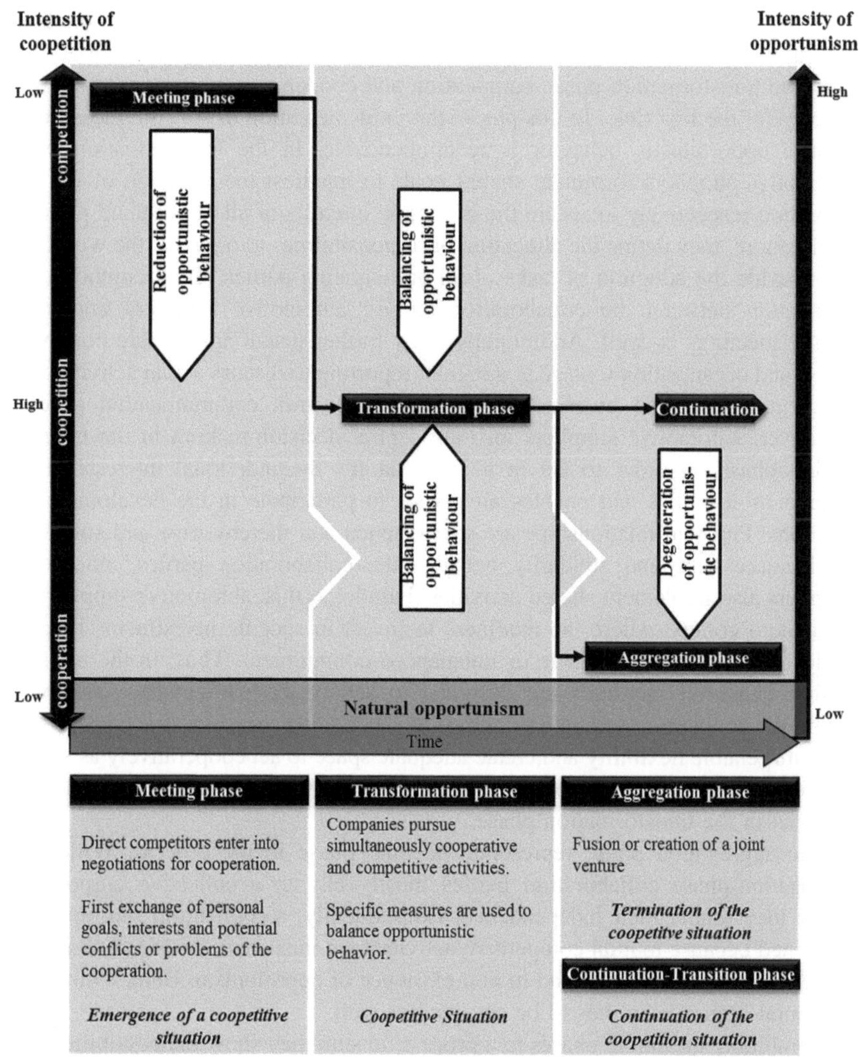


**Fig. 3** Balancing opportunistic behavior in coopetitive settings. *Source* Own illustration

not be influenced directly in this phase, but automotive suppliers can abandon a potential collaboration, if no acceptable severity of opportunism is notable. In this respect the meeting phase secures collaboration parties from choosing inexpedient partners, by establishing trust, solidarity and willingness to undertake obligations as well as interest of a common benefit. If the meeting phase indicates that collaboration parties are well matched and a coopetition setting seems possible, automotive suppliers intensify the collaboration (in critical areas) in the transformation phase. In contrast, if the meeting phase reveals a missing harmony between the collaboration parties or shows that interests and goals are divergent, the meeting phase gives the possibility for an early drop-out (and thus the end of the coopetition setting) without sustaining casualties of substantial length.

In the transformation phase competition and cooperation are practiced simultaneously for the first time. In this phase, the implementation of specific measures to balance opportunistic behavior is recommendable. In the transformation phase automotive suppliers formulate shared goals to manifest the direction of the collaboration respectively to secure the economic interests of all represented partners. Furthermore, they define the allocation of responsibilities to organize the workforce and provide the adoption of tasks of all participating parties. To strengthen communication between the collaboration parties, automotive suppliers implement regular meetings as well. Additionally, they further install appropriate communication and organization tools, e.g. a regular reporting to discuss actual activities and to integrate involved business levels into the overall communication process. Moreover, automotive suppliers institute central decision-making in the transformation phase, in order to act in a way that fits the individual interests of the collaboration parties and enables all parties to participate in the development of solutions. Personal relationships are strengthened and thereby trust and solidarity. To increase trust and solidarity between the collaboration parties, automotive suppliers also implement shared activities. Parallel to this, automotive suppliers try to create a context where the readiness to invest in specific investments helps to avoid feelings of disadvantage or unbalanced commitment. Thus, in the transformation phase all measures are designed to satisfy the participating companies within the coopetition setting. In contrast to sanctioning measures like contracts, all measures enable flexibility and create adequate space to act cooperatively as well as competitively. As in the meeting phase, non-effective measures can also lead to a drop-out in the transformation phase.

The aggregation phase represents the final phase in the process. Within the aggregation-phase collaboration parties mostly ally to a collective corporation, where they render their independence. Consequently, a coopetition setting is discontinued because mutual competitive activities are inexistent. As Fig. 3 shows, an aggregation phase will not lead to non-existence of opportunism. Here, a minimum of natural opportunism has to be accepted as well.

Combining the single phases to a process, in sum they show the possibility for a controllable process of balancing opportunistic behavior in coopetition settings. By doing so, a successful meeting phase succeeds in the transformation phase, enabling the collaboration parties to realize coopetition. As the transformation phase maximizes role ambiguities, the balancing of opportunism herein requires considerable efforts through different measures. To tone down role ambiguities, measures in the aggregation phase reduce competition, whereas the effort of balancing opportunism will be reduced. But the transformation phase and the aggregation phase are not obligatory. They rather afford the opportunity to intensify a collaboration, which is one alternative besides retention or a drop-out. Our cases show that after every phase managerial decisions with awareness of the common past in combination with updated appropriate goals are necessary to ensure long-term benefits of the coopetition setting.

Nevertheless, our results also reveal that in all three phases (especially in the aggregation-phase) opportunism is not completely obstructed. None of the above-mentioned measures are able to fully minimize opportunistic behavior respectively to ensure that the collaboration parties do not have any kind of opportunistic tendencies. According to that, in all three phases collaboration parties have to accept a minimum level of opportunism, which is natural and not preventable for any individual and thus for organizations as a function of individuals.

## 6 Discussion and Conclusion

Our study of coopetition settings in the German automotive supply industry reveals that while in the past managerial attempts aimed at a reduction or elimination of opportunism, in times of coopetition balancing opportunism can be successful to simultaneously achieve cooperative and competitive advantages. Furthermore managers also have to take into account that a minimum level of opportunism must be accepted, since in coopetition settings—as distinguished from strict cooperation—opportunism is important to ensure competitive maneuvers. A possible solution to avoid conflicts in coopetition settings could be the formulation of an acceptable degree of opportunism as a behavioral norm (Hawkins et al. 2008). The unavoidable necessity to accept a low level of natural opportunism and the need of opportunism balancing measures indicates the relevance of rethinking the handling of opportunism in the specific conditions of coopetition. For this purpose our study explored and permitted first insights into this research including a process model as well as measures to balance opportunism.

German automotive suppliers create a context, which on the one hand minimizes opportunism and thereby allows cooperative behavior, but on the other hand also allows clearances for opportunistic behavior and generates opportunities for automotive suppliers to pursue competitive activities. In doing so, the context is mainly guaranteed by relational factors such as trust, solidarity and readiness for comprehensive communication. The automotive suppliers additionally used centralized and formalized structures, a continuous reporting and incentive systems in order to limit strong opportunistic tendencies, while formal contracts or contract-based sanctions are not implemented as measures. Beyond, the presented examples from the automotive supply industry show that specific bilateral investments and equity investments are constructive to ensure a coopetitive context.

Although a balance of opportunistic behavior and thus a coopetitive context is associated with challenges for the automotive industry, coopetition leads to a wide range of advantages. The considered cases show that automotive suppliers were able to achieve economic benefits in terms of a product range expansion, the acquisition of new customer groups, an easier market access, the reduction of overcapacities through the use of production lines of the competitors, decreasing R&D-spending and a higher number of patented innovations. Especially in the transition to electric mobility collaborations with direct competitors and therefore coopetition can play an important role. As the technical and economic challenges of the electric mobility cannot be solved through single automotive suppliers or manufacturers, automotive companies should combine their resources and competencies with direct competitors using coopetitive behavior in order to guarantee adequate economies of scale and scope, to reduce R&D-costs and to share supply-side risks in the transition to electric cars. In this respect, the rethinking of many automotive companies to coopetitive thinking and acting could make a contribution to decreasing sales prices of electric vehicles and thus to an accelerated diffusion of electric mobility.

## References

Achrol RS, Grundlach GT (1999) Legal and social safeguards against opportunism in exchange. J Retail 75(1):107–124

Baiman S (1990) Agency research in managerial accounting: a second look. Acc Organ Soc 15 (4):341–371

Baker GP, Jensen MC, Murphy KJ (1988) Compensation and incentives: practice vs. theory. J Finance 43(3):593–616

Bengtsson M, Kock S (2000) Coopetition in business networks: to cooperate and compete simultaneously. Ind Mark Manag 29:411–426

Bergen M, Dutta S, Walker OC Jr (1992) Agency relationships in marketing: a review of the implications and applications of agency and related theories. J Market 56(3):1–24

Blumberg BF (2001) Cooperation contracts between embedded firms. Organ Stud 22(5):825–852

Bonel E, Rocco E (2007) Coopeting to survive. Surviv Coopetition Int Stud Manag Organ 37 (2):70–96

Brandenburger AM, Nalebuff BJ (1996) Co-opetition. Doubleday, New York

Carson SJ, Madhok A, Wu T (2006) Uncertainty, opportunism, and governance: the effects of volatility and ambiguity on formal and relational contracting. Acad Manag J 49(5):1058–1077

Crosno JL, Dahlstrom R (2008) A meta-analytic review of opportunism in exchange relationships. J Acad Mark Sci 36:191–201

Das TK, Rahmann N (2010) Determinants of partner opportunism in strategic alliances: a conceptual framework. J Bus Psychol 25(1):55–74

Das TK, Teng B-S (1998) Between Trust and control: developing confidence in partner cooperation in alliances. Acad Manag Rev 23(3):491–512

Deeds DL, Hill ChWL (1999) An examination of opportunistic action within research alliances. J Bus Ventur 14:141–163

Dowling MJ, Roering WD, Carlin BA, Wisnieski J (1996) multifaced relationships under coopetition: description and theory. J Manag Inq 5:155–167

Eisenhardt KM (1989) Agency theory: an assessment and review. Acad Manag Rev 14(1):57–74

Eisenhardt KM, Graebner ME (2007) Theory building from cases: opportunities and challenges. Acad Manag J 50(1):25–32

Fojcik TM, Schilling L (2013) Balancing opportunistic behavior in times of coopetition: a multiple case study research of German automotive suppliers. Working paper, Duisburg

Granovetter M (1985) Economic action and social structure: the problem of embeddedness. Am J Sociol 91(3):481–510

Grundlach GT, Achrol RS, Mentzer JT (1995) The structure of commitment in exchange. J Market 59(1):78–92

Hawkins TG, Wittmann CM, Beyerlein MM (2008) Antecedents and consequences of opportunism in buyer-supplier relations: research synthesis and new frontiers. Ind Mark Manag 37:895–909

Heide JB, John G (1992) Do norms matter in marketing relationships? J Market 56(2):32–44
Holmlund M, Kock S (1998) Relationships and the internationalisation of finnish small and medium-sized companies. Int Small Bus J 16(4):46–63
Jeffries L, Reed R (2000) Trust and adaption in relational contracting. Acad Manag Rev 25 (4):873–882
John G (1984) An empirical investigation of some antecedents of opportunism in a marketing channel. J Mark Res 21(3):278–289
Kotzab H, Teller Ch (2003) Value-adding partnerships and coopetition models in the grocery industry. Int J Phys Distrib Logist Manag 33(3):268–281
Luo Y (2007) An integrated anti-opportunism system in international exchange. J Int Bus Stud 38 (6):855–877
Macneil IR (1980) The new social contract: an inquiry into modern contractual relations. Yale University Press, New Haven
Mariani MM (2007) Coopetition as an emergent strategy empirical evidence from an italian consortium of opera houses. Int Stud Manag Organ 37(2):97–126
Milgrom P, Roberts J (1992) Economics, organization and management. Prentice-Hall, Englewood Cliffs
Mysen T, Svensson G, Payan JM (2011) The key role of opportunism in business relationships. Market Intell Plan 29(4):436–449
Nault BR, Tyagi RK (2001) Implementable mechanisms to coordinate horizontal alliances. Manag Sci 47(6):787–799
Poole MS, Van De Ven A (1989) Using paradox to build management and organizational theory. Acad Manag Rev 14(4):562–578
Poppo L, Zenger T (2002) Do formal contracts and relational governance function as substitutes or complements? Strateg Manag J 23(8):707–725
Proff H et al (2012) Competitiveness of the EU automotive industry in electric vehicles. Final Report Framework Contract ENTR/2009/030, European Commission, Brussels
Provan KG, Skinner SJ (1989) Interorganizational dependence and control as predictors of opportunism in dealer-supplier relations. Acad Manag J 32(1):202–212
Rokkan AI, Heide JB, Wathne KH (2003) Specific investments in marketing relationships: expropriation and bonding effects. J Mark Res 40(2):210–224
Walley K (2007) Coopetition: an introduction to the subject and agenda for research. Int Stud Manag Organ 37(2):11–31
Wathne KH, Heide JB (2000) Opportunism in interfirm relationships: forms, outcomes, and solutions. J Market 64(4):36–51
Wilhelm MM (2011) Managing coopetition through horizontal supply chain relations: linking dyadic and network levels of analysis. J Oper Manag 29(7–8):663–676
Williamson OE (1979) Transaction-cost economics: the governance of contractual relations. J Law Econ 22(2):233–261
Williamson OE (1981) The economics of organization: the transaction cost approach. Am J Sociol 87(3):548–557
Williamson OE (1990) Die ökonomischen Institutionen des Kapitalismus: Unternehmen, Märkte, Kooperationen, Tübingen

## Author Biographies

**Lukas Schilling** is Research Associate at the Chair for General Business Administration and International Automotive Management at the University of Duisburg-Essen (Germany). He studied Industrial Engineering and "Automotive Engineering and Management" at the University of

Duisburg-Essen (Germany). His research focuses on Strategic Management, Automotive Management and Change Management.

**Thomas M. Fojcik** is Research Associate at the Chair for General Business Administration and International Automotive Management at the University of Duisburg-Essen (Germany). He studied Business Administration at the University of Hamburg and "Corporate Management and Economics" at the Zeppelin University in Friedrichshafen (Germany). His research priorities are Strategic Management, Innovation Management and Organization in the Automotive Industry.

**Konrad Wirmer** is Research Assistant at the Chair for General Business Administration and International Automotive Management at the University of Duisburg- Essen (Germany). He is studying Industrial Engineering "Power and Management" at the University of Duisburg-Essen (Germany) and studied abroad at the Engineering School "ICAM" in Toulouse (France).

# Valuing Battery Degradation from a Battery Lessor's Perspective

**Björn Hildebrandt, Sebastian Busse, Simon Trang and Lutz M. Kolbe**

**Abstract** In order to promote the market-introduction phase of electric mobility, car manufacturers will have to face the residual value risk of electric cars for consumers, particularly concerning the battery. An increasingly common trend to minimize uncertainties for the consumer, is the separate leasing of the battery. In this paper we examine degradation in lithium-ion batteries from a lessor's perspective. We suppose that battery life can be influenced directly by the user. In the case of a lease contract, the battery always remains the property of the lessor. In this respect, there are no direct incentives for the user to increase battery life by expending additional effort. Therefore, we discuss the idea of a dynamic leasing concept as an opportunity for car manufacturers to encourage consumers to adopt a gentle treatment of the battery in order to extend battery lifetime and thus the residual value. Therefore, we examine the effects of different charging strategies on a battery's residual value to verify whether there is potential for a deeper investigation of a dynamic battery leasing approach. In order to quantify the extent of user influence on battery life, lifetime simulations are performed for different charging scenarios. For our calculations we use a degradation model as a function of the depth of discharge. Furthermore, realistic driving patterns are taken into account to estimate the quantity of energy processed as a function of use mode.

**Keywords** Battery leasing · Degradation · Dynamic pricing · Charging strategy

B. Hildebrandt (✉) · S. Busse
Sustainable Mobility Research Group, University of Göttingen, Göttingen 37073, Germany
e-mail: bhildeb@uni-goettingen.de

S. Busse
e-mail: sbusse@uni-goettingen.de

S. Trang
DFG Research Training Group 1703, University of Göttingen, Göttingen 37073, Germany
e-mail: strang@uni-goettingen.de

L.M. Kolbe
Chair of Information Management, University of Göttingen, Göttingen 37073, Germany
e-mail: lkolbe@uni-goettingen.de



D. Fornahl and M. Hülsmann (eds.), *Markets and Policy Measures in the Evolution of Electric Mobility*, Lecture Notes in Mobility,
DOI 10.1007/978-3-319-24229-3_10

# 1 Introduction

In the Progress Report of the Federal Government's National Platform for Electric Mobility (*Nationale Plattform für Elektromobilität*), it is envisioned that by 2020, one million electrically powered vehicles will be travelling on German roads (NPE 2012). Currently, however, car manufacturers face a low demand for electric vehicles. The primary reasons for this are the high acquisition costs and the uncertainties surrounding the new technology, particularly concerning the battery. Depending on the vehicle type, costs of 10,000–15,000 € for the battery are not uncommon, reflecting a large share of the total acquisition costs.

LeaseTrend head Gerhard Fischer is extremely skeptical of battery electric vehicles (BEV) such as the BMW i3. In a recent issue of "WirtschaftsWoche," he dampened the expectation of the electric car's breakthrough in the fleet business. In Fischer's opinion, the residual value risk is too high; therefore, he urges manufacturers to undertake the risk in order to successfully promote the market introduction phase of electric mobility. Manufacturers can assess the technical risk much better because they have developed the drivetrain and know exactly what can be expected of the batteries: "Honestly, and despite the fact that it is a beautiful car, I cannot recommend anybody to get this car. (…) No one can reliably tell us today how much capacity is left in the battery after 2 or 3 years in everyday traffic. And the battery is the biggest pool of costs in the vehicle" (Rother 2013).

For the widespread use of battery electric vehicles, it is therefore essential to increase the sales of electric cars in order to reduce the costs of lithium-ion batteries through economies of scale. To accomplish this, car manufacturers will have to address consumers in the short term by offering cross-subsidies, extended warranties, or battery leasing. An increasingly common trend in practice is the financial separation of the battery and the vehicle to counteract the cost and residual value problems of electric cars (Diehlmann and Häcker 2012). In such a scenario, the consumer buys or leases the car without power supply and leases or rents the battery separately. Thus, the aim is to minimize uncertainties for the customer, thereby increasing the attractiveness of electric vehicles. Aside from the advantage of lower acquisition costs, the option of leasing the battery separately transfers the financial risk to the lessor. Consequently, the lessor retains the risk of progressive degradation in lithium-ion batteries, which has a direct impact on the residual value of the battery. Although existing batteries have, according to current knowledge, at least 80 % of their power remaining after 10 years and several thousand charge cycles, an uncertainty exists due to a lack of long-term experience. Therefore, the lessor will include the risk of the customer stressing the battery more than average in the leasing rate.

This work considers dynamic battery leasing as an opportunity for car manufacturers to encourage consumers to adopt a gentle treatment of the battery in order to extend battery lifetime and thus the residual value. Longer battery life results in lower leasing rates, which in turn increases the attractiveness of electric cars. In order to enable a financial assessment of the battery-leasing concept, the extent to which a gentle treatment of the battery can increase battery lifetime, and therefore the residual value, is examined. For the purpose of achieving this research objective, specific research questions are formulated and processed in this work:

- RQ1: *Which factors that can be influenced by the user lead to excessive aging of lithium-ion batteries?*
- RQ2: *How large are the effects of different charging strategies on total battery degradation costs?*

## 2 Related Work

There are a variety of studies focusing on aging in lithium-ion batteries. However, a large number of these studies focus on lithium-ion batteries with less capacity, mostly from the consumer sector (e.g., Ramadass et al. 2004; Thomas et al. 2003; Jin et al. 2009; Ning et al. 2003; Liu et al. 2012). For these applications, the requirements are completely different than for battery systems used in electric vehicles or plug-in hybrids. Therefore, the following remarks are limited to studies that focus explicitly on electric mobility applications.

Within the survey of prior literature, several factors that affect the aging of lithium-ion batteries could be identified, including state of charge (SOC), depth of discharge (DOD), temperature, current, and energy processed. The various investigations concentrate primarily on different degradation mechanisms at the outset. Therefore, different model structures and assumptions are used. Focusing on a single subset of existing degradation mechanisms leads to various different estimation models, e.g., as a function of DOD (e.g., Haubrock 2011; Link et al. 2010; Justpark 2011), energy processed (e.g., Peterson et al. 2010) or temperature (e.g., Yuksel and Michalek 2012). Besides, variations are being observed in terms of cell chemistries and cell design. These variations specify the inherent characteristics of the battery and have a significant influence on how the various factors effect degradation phenomena (Yuksel and Michalek 2012). Hence, determining battery degradation is a complex process and there is no single degradation model that is universally valid. This is confirmed by Wenzl et al. (2013) who point out that the interaction of aging mechanisms under complex conditions of use is not fully understood.

Some of the works surveyed within the literature review focus solely on battery degradation phenomena, while others additionally derive algorithms for charging strategies. The authors assume that the object of optimization is a trade-off between minimizing degradation costs and the energy consumption costs of variable electricity tariffs. This implies that it is in the user's interest to minimize the aging costs of the battery. However, an adoption of this assumption does not seem to be worthwhile with regard to the underlying research topic, battery leasing. In the case of a lease contract, the leasing object—in this instance, the battery—always remains the property of the lessor. In this respect, there are no direct incentives for the user to increase battery lifetime by expending additional effort.

Table 1 summarizes the objects that are analyzed in related work and exposes the research contribution of this work. The main goal of this paper is to extend the research mentioned and transfer the findings to a battery lessor's perspective. The dynamic leasing concept aims to increase battery life by prompting users to adopt battery-friendly behavior. In contrast to previous work, we did not focus on the trade-off between minimizing degradation and energy consumption costs; these two optimization approaches pursue different objectives and, from a lessor's perspective, only the aging costs are of interest. For this reason, we examined the effects of different charging strategies on a battery's residual value in order to verify whether there is potential for a deeper investigation of a dynamic battery leasing approach.

# 3 Method of Investigation

## *3.1 Methodological Process*

The procedure of this work is structured as follows. Based on the scientific literature, degradation phenomena in lithium-ion batteries are investigated and conclusions about how the user can influence battery life are drawn. In this context,

**Table 1** Related work and research contribution

| Author | Degradation | Charging strategies | | Residual value consideration from a lessor perspective |
|---|---|---|---|---|
| | | Focus: energy costs | Focus: degradation | |
| Haifeng et al. (2009) | X | – | – | – |
| Bashash et al. (2010) | X | X | X | – |
| Link et al. (2010) | X | X | X | – |
| Peterson et al. (2010) | X | – | – | – |
| Haubrock (2011) | X | – | – | – |
| Hoke et al. (2011) | X | X | X | – |
| Justpark (2011) | X | – | – | – |
| Lunz et al. (2011) | X | X | X | – |
| Moura et al. (2012) | X | X | X | – |
| Yuksel/Michalek (2012) | X | – | – | – |
| Wenzl et al. (2013) | X | – | – | – |
| Research contribution | X | – | X | X |

RQ1 is addressed. Moreover, a degradation model is derived. Including a realistic driving pattern as a further input variable, the model is then used for battery lifetime simulations to quantify the effects of different charging strategies on total battery degradation. Based on these results, implications for the residual value are drawn in order to answer RQ2. Figure 1 summarizes the structure of this work.

## 3.2 Battery Degradation and User Influence

Since the quality of the model being estimated is highly dependent on the quality of the research conducted, a wide-ranging, exploratory research was followed by an extensive, systematic literature review. Despite the existence of many other mechanisms, the primary cause for degradation in lithium-ion batteries is an anode-side reaction. This reaction results in an irreversible formation of the solid electrolyte interface film in the anode, which causes loss of capacity and increases the internal resistance of the battery (Bashash et al. 2010; Moura et al. 2012; Wenzl et al. 2013).

The overall aging process of lithium-ion batteries can be subdivided into calendar life (in standby mode) and cycle life (charging and discharging) (Lunz et al. 2011; Yuksel and Michalek 2012). Wenzl et al. (2013) point out that cycle life and calendar life cause different types of degradation. Depending on the specific application, end of life (EOL) can be either a result of capacity loss (equivalent to driving range) or an increase of internal resistance (equivalent to maximum power

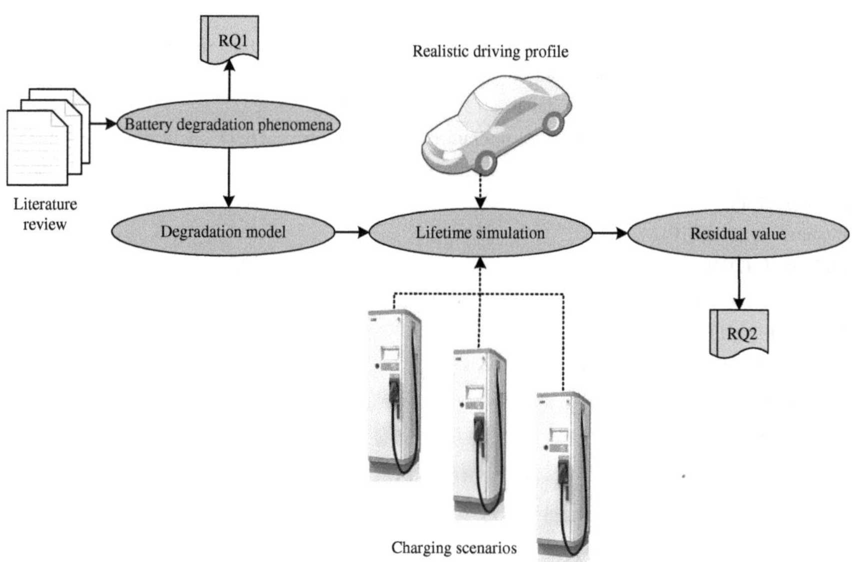


**Fig. 1** Methodological process

capability) (Wenzl et al. 2013). After Yuksel and Michalek (2012) and Shiau et al. (2010), the EOL criterion is defined as the point in time at which the battery has lost 20 % of its capacity or the internal resistance has increased by 30 %, whichever occurs first.

The proceedings reveal that existing scientific literature does not come to a consensus on the different aging mechanisms of lithium-ion batteries. For example, the depth of discharge is discussed as one of the most relevant degradation mechanisms (e.g., Haifeng et al. 2009; Link et al. 2010; Haubrock 2011; Hoke et al. 2011; Lunz et al. 2011; Wenzl et al. 2013), but Peterson et al. (2010) conclude that this has no significant influence on the aging processes. However, it must be noted that they observed the influence of the DOD on aging under relatively low currents. A glance into practice also indicates that no unanimous opinion exists. On the one hand, manufacturers such as Johnson Controls, Nissan, Mitsubishi, Ford, GM, SB-LiMotive, and Toyota state that fast charging and high depths of discharge accelerate the degradation in battery cells. On the other hand, IBM, BYD, Toshiba, and Subaru try to prove the opposite (Justpark 2011).

Table 2 summarizes the findings of the literature studied. It should be noted at this point that the table provides only an overview of the factors that have been analyzed and found to be significant for battery degradation. The absence of an "x" in the table does not necessarily mean that this factor has no effect on aging. This is due to the different experimental models. For example, Peterson et al. (2010) consciously chose the experimental setup with a constant temperature of 25 °C in order to examine the other influences independent of the of temperature.

**Table 2** Identification of degradation factors

| Author | Factors that were analyzed and found to be significant for battery degradation | | | | | |
|---|---|---|---|---|---|---|
| | Calendar life | | Cycle life | | | |
| | SOC | Temperature | Temperature | DOD | Current | Energy processed |
| Haifeng et al. (2009) | – | X | X | X | X | – |
| Bashash et al. (2010) | X | X | – | – | X | – |
| Link et al. (2010) | – | – | – | X | – | – |
| Peterson et al. (2010) | – | – | – | – | – | X |
| Haubrock (2011) | X | X | X | X | X | – |
| Hoke et al. (2011) | X | X | X | X | – | – |
| Justpark (2011) | – | X | X | X | X | – |
| Lunz et al. (2011) | X | X | – | X | – | – |
| Moura et al. (2012) | X | X | – | – | X | X |
| Yuksel/Michalek (2012) | – | X | – | – | – | – |
| Wenzl et al. (2013) | – | X | – | X | X | X |

As the overview shown in Table 2 clearly illustrates, in the case of calendar life, degradation is affected by current SOC and temperature, whereas in the case of cycle life, the factors DOD, temperature, current, and energy processed play an important role. Scientific literature concerning the aging of lithium-ion batteries reveals that battery lifetime can be strongly influenced by operating conditions. Therefore, user behavior can significantly contribute to increasing battery life and thus the residual value of the battery. Focusing on the charging behavior of the users, we further analyze the degree to which user behavior can optimize battery life.

According to the Federal Ministry of Transport, Construction, and Urban Development, a mobile person in Germany makes 3.8 trips/day, travelling a total of 44 km with an average of 11.5 km/trip (Bundesministerium 2008). This implies that the vehicle is not in use for most of the day. Thus, charging management in particular poses a great potential for reducing degradation effects. This hypothesis can be supported by numerous papers dealing with optimization of the charging of electric vehicles (e.g., Bashash et al. 2010; Link et al. 2010; Hoke et al. 2011; Lunz et al. 2011; Moura et al. 2012).

The scientific literature studied indicates that during standby mode (zero or very little current), both the temperature and the SOC have a decisive influence on degradation mechanisms (Bashash et al. 2010; Haubrock 2011; Moura et al. 2012; Yuksel and Michalek 2012; Wenzl et al. 2013). Due to the self-discharge and the resulting threat of deep discharges, lithium-ion batteries should be stored at mid-level SOC. However, high charge levels and the associated high cell voltages accelerate the aging process of the cell (Haubrock 2011). One way of counteracting the calendar aging is the use of fully automated charge controls and, regarding the long-term perspective, the integration of the electric vehicles into the smart grid. By using an intelligent charge control, the time in which the battery is at high SOCs can be minimized. According to recent developments, it is assumed that in the near future, electric vehicles will be charged primarily at the domestic socket, which means that fully automated charging and bidirectional vehicle-to-grid integration cannot be taken into consideration for the time being. Therefore, the vehicle owner theoretically has the possibility of delaying the charging process as long as possible in order to reduce the time that the battery remains at high SOCs, which makes particular sense with long periods of standby mode, e.g., an upcoming holiday. On the other hand, as a vehicle primarily serves mobility demands, a high SOC is desirable to offer as much mobile flexibility as possible. Focusing on the mobility needs, range anxiety is a problem that has a much greater impact than the real range of electric vehicles (Sonnenschein 2010; Eisel and Schmidt 2014). It is assumed that the maximum possible mobile flexibility is a top priority. Therefore, lowering the SOC by not charging the battery despite having the possibility to do so does not seem to be a feasible alternative.

With the thermal effect on battery degradation, it makes sense to actively cool the battery, at least when it is connected to the grid. Nevertheless, today's battery- and temperature-management systems take little account of calendar age. Considering the extensive time that the vehicle spends in standby mode, there is an urgent need to catch up with here (Haubrock 2011). It can be concluded that, in the case of calendar life, primarily technical development can counteract the degradation effects.

Representing the prevailing view of the battery and automobile manufacturers, Johnson Controls affirms a longer life for batteries charged frequently with small charging currents (Justpark 2011). These findings are confirmed by a variety of scientific investigations focusing on degradation in lithium-ion batteries.[1] Electric or plug-in hybrid vehicles will in most cases be charged at a fixed parking space at home or at work. Since the vehicles are usually parked there for several hours, single-phase alternating current charging at 2.3 or 3.6 kW is sufficient (Justpark 2011). The alternative, fast charging, implies a substantially higher stress for the battery. The forgoing of fast charging in everyday life reduces degradation caused by high charging currents.

Moreover, it can be concluded that the user can increase battery life by charging as frequently as possible. By regularly connecting the vehicle to the domestic grid when the vehicle is parked in the garage or in the parking spot, users can significantly reduce the cycle depth. Assuming a corresponding infrastructure—for example, charging stations at work or in public areas—a further reduction of the cycle depths can be achieved. Furthermore, both, small cycle depths and low currents cause lower heating of the cells and therefore reduce temperature-induced degradation (Justpark 2011).

### 3.3 Degradation Model

An estimation model is being developed to predict the resulting battery degradation. However, the objective of the model is neither an exact representation of all degradation effects nor a detailed state-of-health forecast; it is a close-to-reality estimation of the aging mechanisms based on the depth of discharge, which can be influenced by charging behavior, as mentioned above. Following Hoke et al. (2011), the following assumptions are adopted: First, each factor is independent of other factors. Second, the effects are independent of the age of the battery itself.

The relationship between the cycle life and depth of discharge can be represented by an exponential function in which the number of cycles decreases exponentially with a higher cycle depth. This relationship is represented by the following function, where the parameters a and b of the exponential function are the ones to be estimated:

$$N_{cycles} = a * DoD^{-b} \tag{1}$$

Based on this function, for different depths of discharge, the corresponding nominal number of charge and discharge cycles can be estimated. The

[1] See also Table 2.

determination of the parameters is based on data published by Haubrock (2011), who uses real measurement data for estimating battery health. In that case, the estimation parameters of the exponential function are $a = 524{,}007$ and $b = 1.355$.

Figure 2 displays the corresponding approximation function for determining the nominal number of cycles until the EOL criterion is reached. Its course of curve illustrates a significant increase in nominal number of cycles resulting from low DODs, particularly below 10 %. In other words, assuming the same amount of energy processed, high cycle depths cause a higher degradation than lower ones. For this reason, many hybrid vehicles are designed so that their rated capacity—the usable capacity—is limited to only 10 % of the existing physical capacity (Herb 2010). As a result larger batteries are required and, in accordance, a lot of battery capacity is equipped, without being utilized (Herb 2010). With regard to the charging strategy of battery electric vehicles and plug-in hybrids, which have a usable capacity that is not limited to such an extent because of their higher range requirements, this findings indicate that the user can significantly increase the lifetime of a battery when preventing high DODs, i.e., charging as often as possible.

In order to determine how much battery life is exhausted over time, function (1) must be transformed. An aggregation of the individual cycles can determine how much lifetime has been depleted prior to the current cycle; the end of life is reached when remaining life becomes zero. Thus, the residual life L after t cycles can be represented as:

$$L_t = 1 - \sum_{i=1}^{t} \frac{1}{a^*DoD_i^{-b}} \tag{2}$$

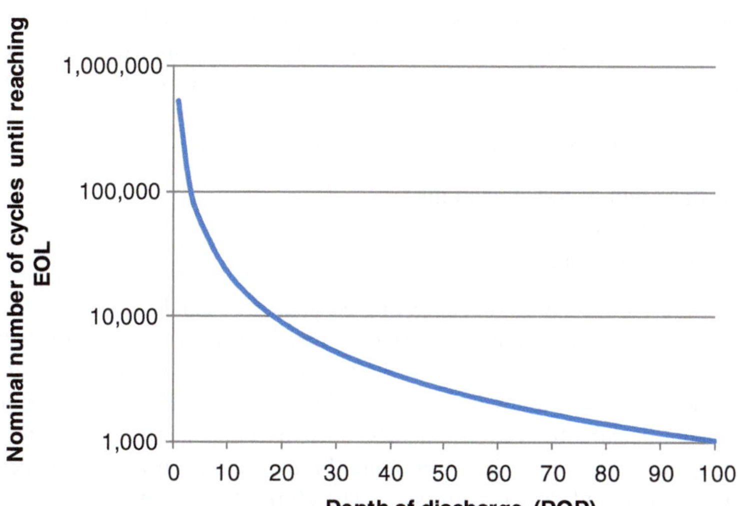


**Fig. 2** Approximation function for determining the nominal number of cycles until reaching the EOL criterion, depending on the depth of discharge

**Table 3** Daily trips

| Start time | End time | Distance driven (km) | Possible reason for trip |
|---|---|---|---|
| 6:30 | 7:00 | 16 | Driving child to school and returning home |
| 12:30 | 14:00 | 22 | Shopping, picking child up from school, and returning home |
| 17:30 | 17:45 | 3 | Short additional shopping trip |

## 3.4 Driving Profile and Charging Scenarios

In order to quantify the effects of different charging strategies on total battery degradation costs, different charging scenarios are simulated. To estimate the quantity of energy processed to and from the battery as a function of use mode, a realistic driving pattern is taken into account. Brandt et al. (2013) construct a rural driving profile based on the study "Mobility in Germany, 2008," which documents the trips of about 26,000 households and over 60,000 individuals. The driving profile considered fits to any household which has one car parked at home for a significant part of the day (Brandt et al. 2013). It consists of three home-to-home tours, which have a total distance of 41 km (Table 3).

Figure 3 visualizes the daily driving profile, which represents all trips taken by the vehicle over the whole day. The horizontal sections of the line imply that the vehicle is parked at home and can be connected for charging, while the diagonal sections indicate that the vehicle is in use.

It is assumed that when the car is parked and connected to the grid, the charging process starts immediately and continues until the maximum SOC is reached or the next tour begins. The calculations are based on the technical data of a VW e–up!, which has a rated battery capacity of 18.7 kWh. A combined consumption in accordance with the new European driving cycle (NEDV) of 11.7 kWh/100 km results in a range of 160 km.[2] Three different scenarios are considered to estimate the degradation effects of different charging strategies:

1. Immediate end-of-travel charging
2. Charging overnight
3. Charging on demand

*Immediate end-of-travel charging* implies that the vehicle is connected to the grid for charging right after arriving home. When *charging overnight*, the battery is charged after the last tour (e.g., at 17:45) each evening. *Charging on demand* ensures mobility at all times with a safety buffer. Actual energy consumption is affected significantly by various factors, such as cold ambient temperatures or electronic accessories, e.g., air conditioning. Furthermore, consumption depends on the type of route and driver aggression. Therefore, it is assumed that the battery

---

[2] *Source* Datasheet of the VW e-up!

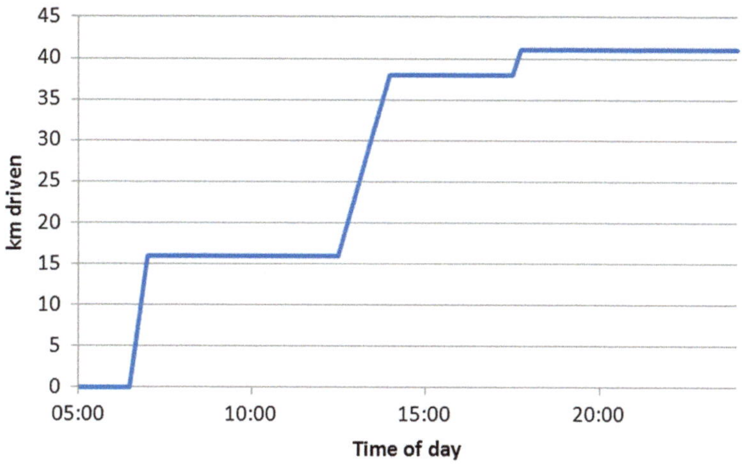


**Fig. 3** Daily driving profile

must be charged after 3 days of use, corresponding to a driven distance of 123 km, which means a safety buffer of 37 km or 23.13 % out of a total range of 160 km.

# 4 Results

## *4.1 Lifetime Simulation*

Following the methodological process depicted in Fig. 1, we computed the remaining life for the three different charging strategies (immediate end-of-travel charging, charging overnight and charging on demand). Thereby, we used function (2) for estimating battery degradation and included the driving profile presented. Figure 4 summarizes the results in a period of one to 10 years of use. It is assumed that only a single charging strategy is applied for the entire time period. A remaining battery life of 100 % indicates a brand new battery, whereas the end of life is reached at 0 %.

It is apparent that *end-of-travel charging* results in significantly higher battery life than the other charging strategies. After each trip the SOC is raised until 100 % to reduce the cycling aging caused by high DODs. This results in a higher number of total cycles and therefore in a longer lifetime referring to the number of driving days. Even after 10 years of use, the remaining battery life is still 58.25 %, whereas *charging on demand* results in just 16.50 % battery life remaining. Taking the same distance driven as a reference point, the savings achieved by the avoidance of deep discharges are comparatively huge. These findings indicate that *end-of-travel charging* results in a battery lifetime of several decades, assuming the daily driving profile of an average citizen.

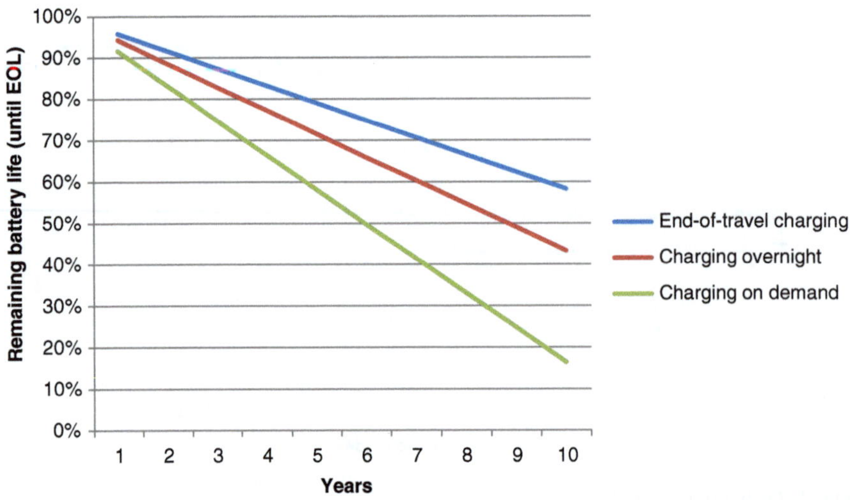


**Fig. 4** Lifetime simulation: remaining battery life as a function of driving profile and charging scenarios

## 4.2 *Implications for the Residual Value*

After the influence of different charging strategies on battery life has been estimated, implications for the residual value are exposed. According to a survey published by McKinsey in June 2012, the current price of a lithium-ion battery system for electric vehicles is priced at 500–600 $/kWh (Hensley et al. 2012).[3] When adopting the figures reported by McKinsey, a 18.7 kWh battery installed in the VW e-up!, would cost between 6,923 and 8,308 €.[4] Applying the mean value, a brand new VW e-up! battery is charged with 7,616 €. Within our calculations, we assume a linear relationship between the remaining life and the residual value, and it is presumed that the residual value is zero when the EOL criterion is reached.[5] Table 4 summarizes the development of the residual value over time for the different charging scenarios. Furthermore, it displays the savings that can be achieved by switching from *charging on demand* to *end-of-travel charging*.

The monetary approach presented in Table 4 exposes that there is great potential for increasing the residual value with battery-friendly charging strategies. In the case of a 36-month lease term, *end-of-travel charging* results in a loss of 953.98 €,

---

[3]These figures apply to the complete battery pack, including battery cells, battery management systems, and casing. The price applies per effective kWh, assuming batteries with 70 % DOD (Hensley et al. 2012).

[4]The conversion is based on www.oanda.com on 07/11/2013. An exchange rate of $ 1 = € 0.74047 is used, with results rounded to integer numbers.

[5]In the case that the battery can be resold following its use in the vehicle, e.g. in the context of a second life concept, this assumption has to be adjusted.

**Table 4** Progress of residual value for different charging strategies

| Residual value | End of travel-charging (€) | Charging over night (€) | Charging on demand (€) | Savings of end-of-travel-charging compared to charging on demand (€) |
|---|---|---|---|---|
| New battery | 7,616.00 | – | – | – |
| After 1 year | 7,298.01 | 7,185.42 | 6,980.04 | 317.96 |
| After 2 years | 6,980.01 | 6,754.85 | 6,344.08 | 635.93 |
| After 3 years | 6,662.02 | 6,324.27 | 5,708.12 | 953.89 |
| After 4 years | 6,344.02 | 5,893.69 | 5,072.16 | 1,271.86 |
| After 5 years | 6,026.03 | 5,463.11 | 4,436.20 | 1,589.82 |
| After 6 years | 5,708.03 | 5,032.54 | 3,800.24 | 1,907.79 |
| After 7 years | 5,390.04 | 4,601.96 | 3,164.28 | 2,225.75 |
| After 8 years | 5,072.04 | 4,171.38 | 2,528.33 | 2,543.72 |
| After 9 years | 4,754.05 | 3,740.80 | 1,892.37 | 2,861.68 |
| After 10 years | 4,436.05 | 3,310.23 | 1,256.41 | 3,179.65 |

which is doubled by *charging on demand* (1,907.88 €). Similar results are obtained at a lease term of 60 months, in which savings of 1,589.82 € can be realized by switching from the strategy *charging on demand* to *end-of-travel charging*.

## 5 Limitations

The following limitations should be considered for interpreting the results. First, as mentioned above, our degradation model does not claim to be universally valid, since there exist various different battery chemistries and designs, which cannot be captured within a single degradation model. We focus on cycle life, assuming that the end of life will be reached as a result of cycle aging before it has been reached due to calendric aging. Calendar life is not taken into account, since the interaction of aging mechanisms under complex conditions of use in electric cars and their impact on performance characteristics are not fully understood (Wenzl et al. 2013). This gap should be addressed by further research to predict the interaction of cycling and calendric aging. If extensive degradation data is available, e.g., on the part of battery manufacturers, this can be used for a more precise estimation. Second, the driving profile considered fits to a household in rural regions, where one car is parked at home for a significant part of the day. The total distance of 41 km driven per day results in a comparatively high battery lifetime. Larger distances driven by vehicles that are used in commercial transportation, e.g., by taxi operators or car-sharing providers result in a much shorter battery lifetime and a stronger decline in value.

## 6 Discussion and Conclusion

In order to promote the market-introduction phase of electric mobility, car manufacturers will have to face the residual value risk of electric cars for consumers, particularly concerning the battery. One opportunity to minimize uncertainties for the consumer is the separate leasing of the battery, which transfers the financial risk to the lessor. In currently existing approaches, the risk of the customer stressing the battery more than average will be included in the leasing rate by the lessor. As an alternative, we discussed the basic idea of adopting a dynamic battery leasing concept. As a first step in assessing this concept, we evaluated the effects of different charging strategies on battery degradation and drew conclusions to the residual value. We then proposed a model for a close-to-reality estimation of realistic degradation mechanisms as a function of the depth of discharge, which is based on real measurement data published by Haubrock (2011). Realistic driving patterns were taken into account to estimate the quantity of energy processed as a function of use mode. In order to quantify the extent of user influence on battery life, lifetime simulations were performed for three different charging scenarios.

The contribution of this work is examining the basic idea of dynamic battery leasing in a monetary context. Addressing RQ1, it was shown that battery life can be strongly influenced by its operating conditions. Especially the charging management poses a great potential for the user to reduce degradation effects by influencing the depth of discharge and the actual state of charge of the battery. Further, the avoidance of fast charging in everyday life reduces degradation caused by high charging currents.

Our calculations show that different charging strategies have a significant impact on battery lifetime. *End-of-travel charging* returns the best results, indicating that charging as often as possible should be favored for increasing battery life. By monetarily assessing degradation costs, we were able to answer RQ2. The financial attractiveness of considering a dynamic battery leasing approach, referring to the actual customer behavior, is thus quite apparent. This paves the way for further research focusing on relevant design principles, developing policies, and the technical implementation for a dynamic battery leasing approach.

## References

Bashash S, Moura SJ, Fathy HK (2010) Charge trajectory optimization of plug-in hybrid electric vehicles for energy cost reduction and battery health enhancement. In: American control conference marriott waterfront, Baltimore, MD, USA, June 30–July 02. IEEE, Piscataway, pp 5824–5831

Brandt T, Feuerriegel S, Neumann D (2013) A household-orientated approach to the benefits of vehicle-to-grid-capable electric vehicles. In: Proceedings of the Multikonferenz Wirtschaftsinformatik 2013, Leipzig, pp 1667–1682

Bundesministerium (2008) Mobilität in Deutschland 2008: Ergebnisbericht: Struktur – Aufkommen – Emissionen – Trends. Berlin

Diehlmann J, Häcker J (2012) Automobilmanagement. Oldenbourg Verlag, München

Eisel M, Schmidt J (2014) The value of IS for increasing the acceptance of electric vehicles–the case of range anxiety. In: Proceedings of the Multikonferenz Wirtschaftsinformatik 2014, Paderborn, pp 882–894

Haubrock A (2011) Degradationsuntersuchungen von Lithium-Ionen Batterien bei deren Einsatz in Elektro- und Hybridfahrzeugen. Cuvillier Verlag, Göttingen

Haifeng D, Xuezhe W, Zechang S (2009) A new SoH prediction concept for the power lithium-ion battery used on HEVs. In: Proceeding of vehicle power and propulsion conference, VPPC 2009, Dearborn, MI, USA, September 7–11, 2009. IEEE, Piscataway, pp 1649–1653

Hensley R, Newman J, Rogers M, Shahinian M (2012) Battery technology charges ahead. McKinsey Quarterly, July 2012, pp 1–9

Herb F (2010) Alterungsmechanismen in Lithium-Ionen-Batterien und PEM-Brennstoffzellen und deren Einfluss auf die Eigenschaften von daraus bestehenden Hybrid-Systemen. Dissertation. Universität Ulm, Ulm

Hoke A, Brissette A, Pratt A, Smith K (2011) Electric vehicle charge optimization including effects of lithium-ion battery degradation. In: Proceeding of vehicle power and propulsion conference, VPPC 2011, Chicago, IL, USA, September 6–9. IEEE, Piscataway

Justpark (2011) Abschlussbericht zum Verbundvorhaben Begleitforschung zum kabellosen Laden von Elektrofahrzeugen. Institut für Automation und Kommunikation e.V. Magdeburg an der Otto-von-Guericke-Universität Magdeburg, Kiefermedia GmbH, Offenburg

Jin HF, Liu Z, Teng YM, Gao JK, Zhao Y (2009) A comparison study of capacity degradation mechanism of LiFePO4-based lithium ion cells. J Power Sources 189:445–448

Link J, Büttner M, Dallinger D, Richter J (2010) Optimisation algorithms for the charge dispatch of plug-in vehicles based on variable tariffs. Working paper sustainability and innovation, No. S3/2010

Liu D, Luo Y, Peng Y, Peng X, Pecht M (2012) Lithium-ion battery remaining useful life estimation based on nonlinear ar model combined with degradation feature. In: Proceedings of annual conference of the prognostics and health management society, PHM 2012, Minneapolis, MN, USA, September 23–27, 2012, pp 24–27

Lunz B, Walz H, Sauer DU (2011) Optimizing vehicle-to-grid charging strategies using genetic algorithms under the consideration of battery aging. In: Proceeding of vehicle power and propulsion conference, VPPC 2011, Chicago, IL, USA, September 6–9. IEEE, Piscataway

Moura SJ, Stein JL, Fathy HK (2012) Battery-health conscious power management in plug-in hybrid electric vehicles via electrochemical modeling and stochastic control. IEEE Trans Control Syst Technol 21(3):679–694

Ning G, Haran B, Popov BN (2003) Capacity fade study of lithium-ion batteries cycled at high discharge rates. J Power Sources 117:160–169

NPE (2012) Fortschrittsbericht der Nationalen Plattform Elektromobilität (Dritter Bericht). Berlin

Peterson SB, Apt J, Whitacre JF (2010) Lithium-ion battery cell degradation resulting from realistic vehicle and vehicle-to-grid utilization. J Power Sources 195:2385–2392

Ramadass P, Haran B, Gomadam PM, White R, Popov BN (2004) Development of first principles capacity fade model for li-ion cells. J Electrochem Soc 151(2):196–203

Rother FW (2013) Interview Lease Trend-Chef "Ich kann das Auto keinem empfehlen". WirtschaftsWoche. http://www.wiwo.de/technologie/auto/leasetrend-chef-ich-kann-das-auto-keinem-empfehlen-seite-all/9079722-all.html. Accessed 18 Nov 2013

Shiau CSN, Kaushal N, Hendrickson CT, Peterson SB, Whitacre JF, Michalek JJ (2010) Optimal plug-in hybrid electric vehicle design and allocation for minimum life cycle cost, petroleum consumption, and greenhouse gas emissions. J Mech Des 132:1–11

Sonnenschein J (2010) Preparing for The Roll-Out of Electric Vehicles: Exploring how cities can become plug-in ready. IIIEE Theses. Lund University, Lund

Thomas EV, Case HL, Doughty DH, Jungst RG, Nagasubramanian G, Roth EP (2003) Accelerated power degradation of Li-ion cells. J Power Sources 124:254–260

Yuksel T, Michalek J (2012) Development of a simulation model to analyze the effect of thermal management on battery life. SAE Technical Paper 2012-01-0671

Wenzl H, Haubrock A, Beck HP (2013) Degradation of lithium ion batteries under complex conditions of use. Int J Res Phys Chem Chem Phys 227(1):57–71

## Author Biographies

**Björn Hildebrandt** is research associate and doctoral student at the Georg-August-University of Göttingen, Germany. Furthermore, he is member of the university's Sustainable Mobility Research Group. Björn graduated with a master's degree in Information Management at the Georg-August-University Göttingen. Before that he studied Information Systems at the Westfälische Wilhelms-Universität Münster, Germany, where he received his bachelor's degree. His current research focuses on the impact of digital technologies on business models in the mobility sector.

**Sebastian Busse** is a postdoc at the Georg-August-University of Göttingen, Germany and is the head of the university's Sustainable Mobility Research Group. Sebastian holds a Masters degree of Information Systems from the Technical University Braunschweig, Germany. His current research focus is the utilization of information technology for sustainable mobility solutions and its feasibility assessment. Sebastian's research has been published in outlets such as the International Conference on Information Systems Proceedings and European Conference on Information Systems proceedings.

**Simon Trang** is research member and Ph.D. candidate at the DFG research training group 1703 at Göttingen University. He studied information systems at Göttingen University, Aarhus University, and Seoul National University and received his master's degree in 2011. His research interests lie at the interface between information systems, strategic management, and ecological sustainability. He has published in leading IS conferences such as European Conference on Information Systems and Hawaii Conference on Systems Sciences.

**Lutz M. Kolbe** Since 2007 Lutz M. Kolbe is a full professor and Chair of Information Management at the Georg-August-University of Göttingen, Germany. Lutz completed an apprenticeship in banking and received his degree in Information Systems from the Technical University Braunschweig, Germany. At the Technical University Freiberg, Germany, he earned his doctorate in Business Administration. Subsequently, he led Deutsche Bank's IT think tanks in Frankfurt, New York, and San Francisco. In 2002 he moved on to the Institute of Information Systems at the University of St. Gallen, Switzerland, as head of several competence centers. He was also a visiting scholar at the University of Rhode Island, New York University and IBM Research Almaden (USA). His main research interests are Green IS, IT Innovation Management and Sustainable Mobility.

# The Influence of Organisational and Structural Parameters of Commercial Fleets on the Substitution Potential of Battery Electric Vehicles

**Heike Flämig, Christian Matt, Christian Rudolph and Sören Christian Trümper**

**Abstract** In this paper 969 commercial fleets in the City of Hamburg, Germany, are analysed to investigate the influence of organisational parameters, like vehicle use, as well as structural parameters, such as vehicle types and fleet size, on the potential of battery electric vehicles to substitute conventional ones. The evaluation of the 8,466 reported vehicles shows a theoretical substitution potential, as 90 % of the reported vehicles were below 3.5 tonnes gross vehicle mass and 75 % were used for trips below 140 km a day as the number of the average maximum range of those BEVs available in the market. For those occasions where the range of a BEV will not suffice, alternative mobility options or transport modes still seem a prerequisite for the deployment of BEVs, especially since 70 % of the investigated fleets had no more than four vehicles and little vehicle redundancy. But, 48 % of the companies do also use other transport modes during working hours such as public transport or bicycles. In addition, vehicle replacement cycles proved to be another influential structural parameter. It is dictating the possible rate of penetration of BEVs in the fleets and limiting the substitution potential of the investigated fleets to 40 % over the next 3 years.

**Keywords** Battery electric vehicles · BEV · Commercial fleets · Substitution potential · Fleet size · Economic sectors · Urban freight

H. Flämig · C. Matt · S.C. Trümper (✉)
Hamburg University of Technology, Institute for Transport Planning and Logistics (W8), Schwarzenbergstrasse 95 (E), 21073 Hamburg, Germany
e-mail: truemper@tuhh.de

H. Flämig
e-mail: flaemig@tuhh.de

C. Matt
e-mail: christian.matt@tuhh.de

C. Rudolph
German Aerospace Center (DLR), Institute for Mobility Research, Rutherfordstrasse 2, 12489 Berlin, Germany
e-mail: christian.rudolph@dlr.de

 
D. Fornahl and M. Hülsmann (eds.), *Markets and Policy Measures in the Evolution of Electric Mobility*, Lecture Notes in Mobility,
DOI 10.1007/978-3-319-24229-3_11

# 1 Introduction

Constant efforts in climate change mitigation and the accompanying legislative measures identify alternative fuels, or drivetrains respectively, as a promising means to reduce emissions and fuel consumption from the transport sector while at the same time sufficiently serving the current mobility demand of Western societies. During the last years, major technological innovations once again brought the battery electric vehicle (BEV) into the focus of transport policies, e.g. as part of the German Federal Development Scheme for Electric Mobility (Federal Republic of Germany 2009). Since the year 2009, for instance, more than 700 BEVs have been deployed within this scheme in the City of Hamburg, Germany's second largest city (Trümper 2013). It is because of these numbers that industry and politics are becoming increasingly curious to whether the market potential of BEVs is large enough for realising the necessary economies of scale in production.

Existing studies already predict a high potential for BEVs in companies (Handelskammer Hamburg 2013) on the grounds of a large vehicle stock and the positive attitudes of decision makers towards BEVs. Being such a promising field of early deployment, commercial fleets might eventually pave the way to acceptable, i.e. lower prices of the vehicles. The goal of this survey is to further qualify the substitution potential of BEVs in commercial fleets. It was conducted by the Hamburg University of Technology in cooperation with the local Chamber of Commerce amongst the members of the chamber. The hypothesis is that organisational and structural parameters of the fleets influence the potential within companies to substitute conventional vehicles in their fleet with BEVs. Organisational parameters describe direct aspects like modal choice or vehicle fleet management as well as indirect aspects such as purchase strategies. Structural parameters on the other hand comprise the vehicle size or the available vehicle types in the fleet or preferred sales areas.

# 2 Sample and Methodology

Between December 2012 and February 2013 more than 46,000 companies listed in the Hamburg register of commercial companies were sent a printed questionnaire consisting of two parts. The first part, encompassing ten questions, asked for general information about the company such as type of business, number of employees, goods and services provided. In the second part, consisting of six questions, the survey asked for more detailed information about the fleet, such as types and ownership of vehicles and how the conventional vehicles are operated, for instance by enquiring about daily vehicle kilometres.

The survey resulted in 1,696 replies. 13 of 21 economic sectors and 27 of 88 industries of the European statistical nomenclature (NACE) were covered. The composition is a random result since all registered companies were addressed in the questionnaire and no pre-selection was made. 717 (43 %) of the responding companies had no vehicle. The remaining 969 (57 %) companies provided information on 8,466 vehicles in commercial use in Hamburg. These companies operate between one and 350 vehicles of different types.

About 80 % of the questionnaires were filled in and returned on paper. As mandatory questions are not possible in paper-and-pencil-surveys and respondents left out different questions, the sample size for each question varied and is given in brackets in each case.

The sample encompassed a broad range of 27 industries listed in the commercial register, from trading over real estate agents, media companies and pharmacies to financial brokers etc. (Table 1).

# 3 Vehicle Types

The companies were asked about the composition of their fleets, grouped into different vehicle types, starting with small (subcompact, compact) and mid-sized cars to more spacious vehicles like large cars, station wagons, vans and trucks below 3.5 tonnes. All vehicles above 3.5 tonnes of gross vehicle mass and special purpose vehicles were covered by the category ‘other vehicles’. The latter is an important distinguishing mark in the context of electric vehicles since currently there are no purely battery-powered vehicles above 3.5 tonnes available as series products on the market.

Figure 1 shows the distribution of vehicle types according to industries. The majority of the industries (56 %) used smaller cars in the subcompact or compact category. The following industries mostly used larger vehicles (cars, vans, trucks and station wagons): hotels, restaurants and catering; passenger transport; finance/insurance/tax services; facility management; nursing; bakery/confectionary; wholesale retailers including B to B; retailers (B to C), retailers with customer service; miscellaneous; engineering services; construction craft as well as meeting/event agencies. 19 industry types used ‘other vehicles’, representing 10 % in total of the vehicles in the sample. In two of the 19 industries covered, the share of other vehicles exceeded 25 % of the fleet: 67 % of the total fleet in the transport/logistics/warehousing industry and 70 % in the meeting and event agency industry were ‘other vehicles’.

The analysis showed a high variability in the vehicle composition of the fleets in the industries covered.

In the sample, 90 % of the vehicles were smaller than 3.5 tonnes, representing vehicles which are available on the market as a BEV-equivalent in respect to size, like number of seats or size of freight space, and payload, which is the weight allowance of the vehicle for freight of any kind.

**Table 1** Overview of the responding companies by economic sector, industries and numbers of vehicles in industries

| Economic sector | No. of companies in economic sector | Industry | No. of companies in industry | No. of vehicles in industry |
|---|---|---|---|---|
| Manufacturing | 51 | Production | 46 | 483 |
| | | Bakery/Confectionery | 5 | 29 |
| Electricity, gas, steam and air conditioning supply | 7 | Energy suppliers and service providers | 7 | 402 |
| Construction | 49 | (Construction-) Craft | 49 | 364 |
| Wholesale and tetail trade, repair of motor vehicles and motorcycles | 162 | Pharmacy | 13 | 24 |
| | | Wholesale retailers including B to B | 86 | 1353 |
| | | Retailers (B to C) | 41 | 318 |
| | | Retailers with customer service | 22 | 197 |
| Transportation and storage | 69 | Passenger transport | 13 | 119 |
| | | Transport/logistics/warehousing | 33 | 562 |
| | | Shipping Co. and | 23 | 246 |
| | | Shipping-related services | | |
| Accomodation and food service activities | 30 | Hotel industry/gastronomy | 15 | 46 |
| | | Delivery Services (food) | 15 | 94 |
| Information and storage | 79 | Publishing industry | 6 | 20 |
| | | IT & ICT services | 73 | 517 |
| Financial and insurance activities | 47 | Finance/insurance/tax services | 47 | 196 |
| Real estate activities | 63 | Real estate and property management | 63 | 532 |
| Professional, scientific and technical activities | 139 | Architecture and planning services | 13 | 30 |
| | | Consulting/advertising agency | 92 | 256 |
| | | Engineering services | 34 | 207 |
| Administrative and support service activities | 29 | Facility management | 8 | 61 |
| | | Security services | 5 | 66 |
| | | Temp. employment Co./ personnel services | 16 | 40 |
| | 12 | Nursing | 12 | 220 |

(continued)

**Table 1** (continued)

| Economic sector | No. of companies in economic sector | Industry | No. of companies in industry | No. of vehicles in industry |
|---|---|---|---|---|
| Human health and social work activities other service activities | 232 | Other services | 178 | 1603 |
| | | Meeting and event agency | 9 | 76 |
| | | Miscellaneous | 45 | 405 |
| Total | | | 969 | 8466 |

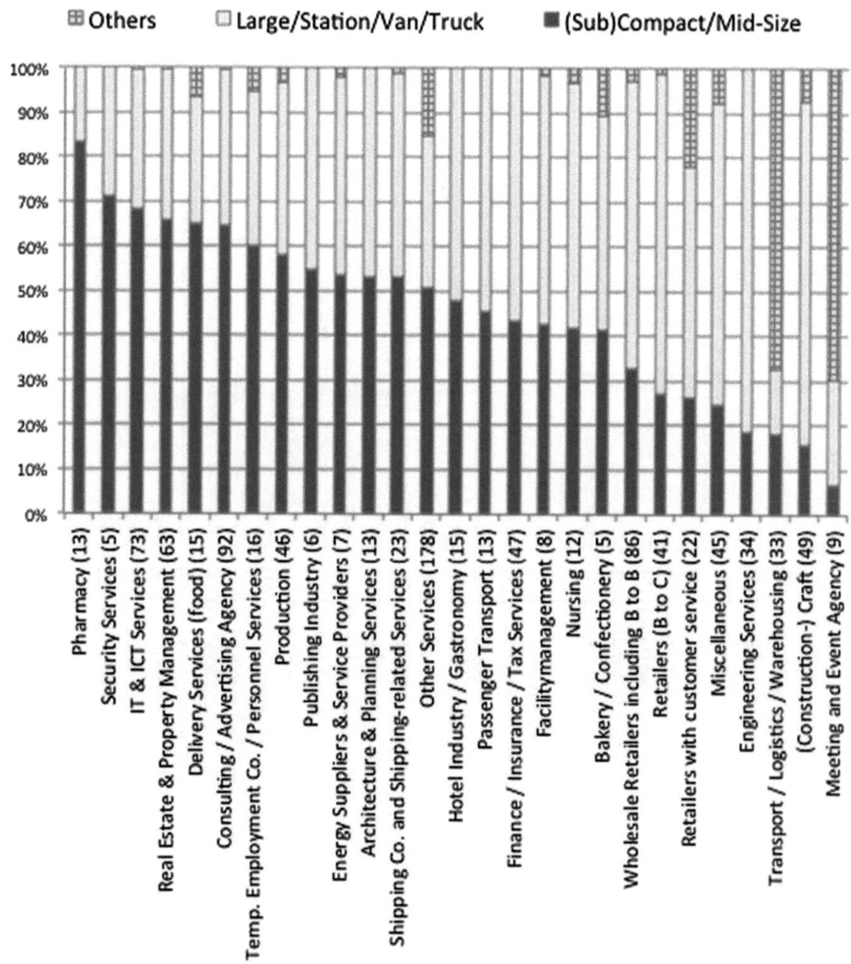


**Fig. 1** Percentage of vehicle types in the fleets of the responding companies grouped by industries and sorted according to the share of smaller vehicles

## 4 Average Daily Vehicle Kilometres

While the analysis of the vehicle types indicates a general substitution possibility for most of the reported vehicles, the key question for the use of BEVs—and one of the main concerns for many users (BMVBS 2011)—is whether their range is sufficient for the daily requirements of their business.

The average daily vehicle kilometres of the reported vehicles were analysed along the categorisation of vehicle types and industries. Anticipating the possible difficulty respondents would have to indicate this figure with precision for each individual vehicle in their fleet, the questionnaire used a redundant "cascade" in which it would, in addition, inquire if the vehicles in question were travelling above or below 140 km a day. The number of 140 km was chosen because at the time

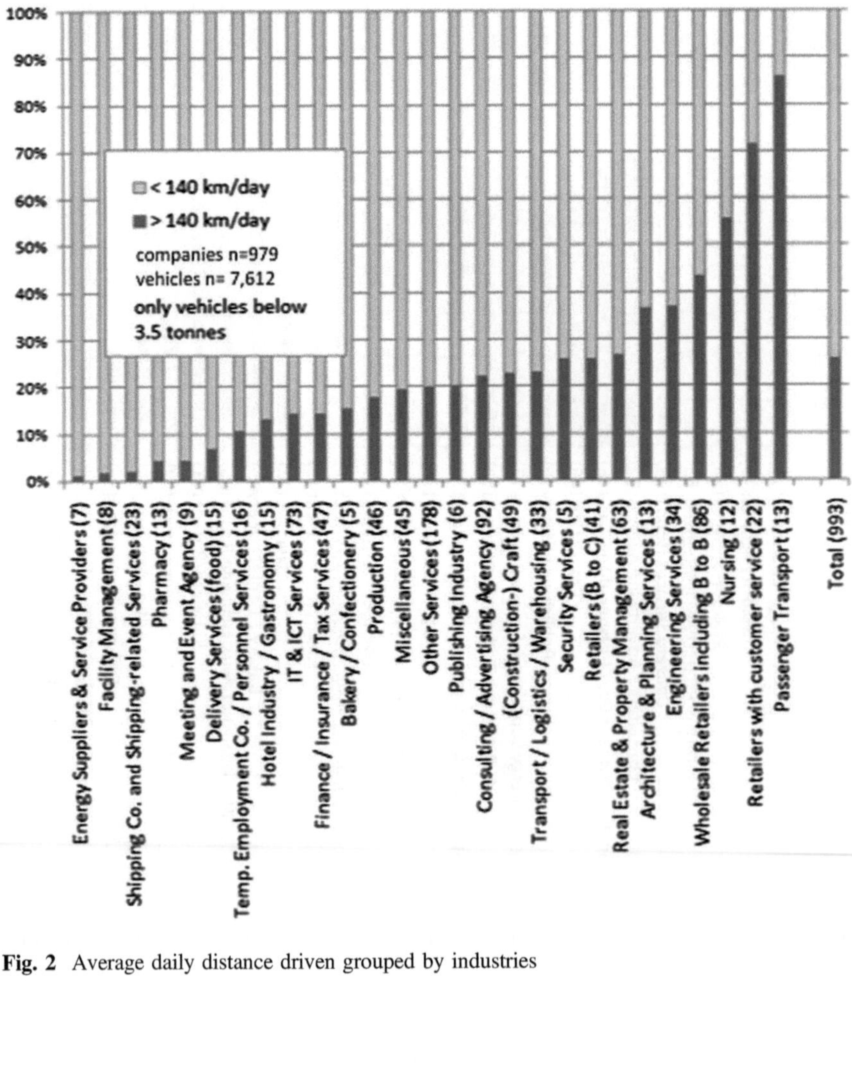


**Fig. 2** Average daily distance driven grouped by industries

when the questionnaire was designed, this figure was regarded as the average maximum range of those BEVs available in the market.

Of the vehicles reported in this survey, 75 % drove less than 140 km a day on average (Fig. 2). This is in line with the general average daily travel distance of commercially used vehicles in Germany which is 72 km/day, in contrast to 36 km/day for vehicles registered to a private person (BMVBS 2012).

It becomes obvious that the average daily travel distance is an ambivalent factor. While a higher daily mileage can generally help to increase the cost-effectiveness of the currently still higher-priced BEVs (Holzer 2014), it also poses a greater risk of running out of battery energy, especially when heavy loads are carried or parts of a trip are covered at higher speeds, e.g. on motorways. Total cost of ownership (TCO) studies indicate a lower annual TCO for BEVs compared to conventional drivetrains (diesel, petrol) and even plug-in hybrid vehicles (PHEV) in the corridor between 10,000 and 45,000 km annual mileage (Gnann et al. 2012). Translated into daily vehicle kilometres, cost-effectiveness of a BEV can be reached, if the daily vehicle kilometres lie between 50 and about 150 km (own calculation based on Gnann et al. 2012), wherein the upper maximum is limited by the maximum range of today's BEVs.

## 5 Fleet Sizes

The fleet sizes were analysed to gain an overview over the possibilities of companies for selecting a specialised vehicle for a specific situation. Or put another way, knowing the fleet size and composition allows a more accurate evaluation of the loss of organisational flexibility due to a BEV.

The histogram in Fig. 3 shows the fleet sizes reported in the survey. One third of all responding companies with vehicles operated just a single vehicle rather than a fleet. About 70 % of all these companies had no more than four vehicles in operation. The largest fleet consisted of 350 vehicles.

Small fleets or single vehicles could be an obstacle to the market penetration of fleets by BEVs because of less flexibility for the fleet management. It can be assumed that smaller fleets have less vehicle-stock redundancy. From an operational perspective even one or two situations a month where a product cannot be delivered or a service cannot by provided to the customer in time or a taxi ride has to be denied on the grounds of an insufficient remaining state of charge is often a plenty of reason against the procurement of a BEV. Thus only 30 % of all responding companies with vehicles appear to have the necessary organisational flexibility in the fleet required for the use of a BEV.

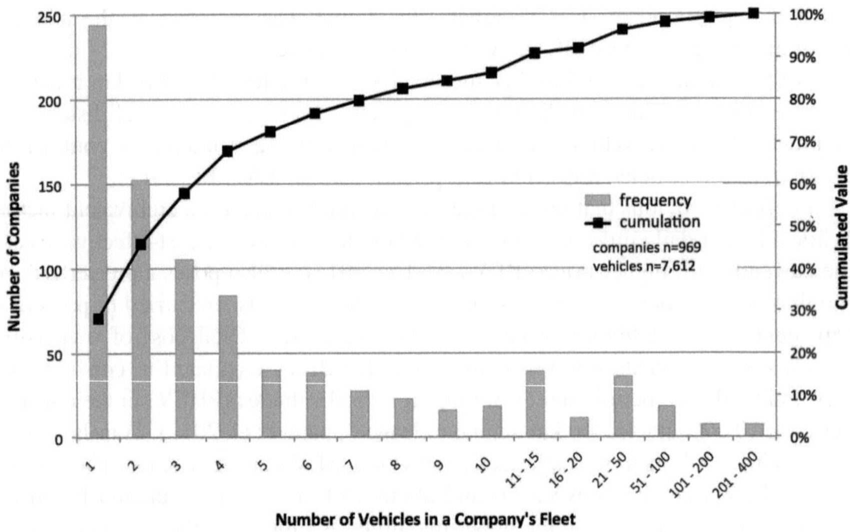


**Fig. 3** Fleet size as interval of vehicle numbers in companies

# 6 Transport Modes Used

Corporate vehicles might not be the only means of transport in the participating organisations. Companies were asked to indicate any other modes of transport they used in carrying out their commercial activities, e.g. using public transport, taking the bicycle or carrying out tasks on foot when covering short distances. The substitution potential could be lowered if the environmental and transport policy strategies supporting the modal shift become effective and the individual, motorised vehicle is substituted by other modes. At the same time the substitution potential could be raised if the environmental and transport policy strategies supporting the modal shift become effective for passenger transport: for long distances the companies could use the train, for short distances the BEV.

Figure 4 shows the share of alternative transport modes used by the responding companies. Of the responding companies 386 (48 %) companies use additional modes of transport, 332 (86 %) use public transport, 210 (54 %) use employee's own vehicles, 161 (42 %) use the bicycle and 123 (32 %) also walk as part of their routines (multiple answers allowed).

The use of other transport modes in a company is thus an indicator for a more flexible mobility management and, if developed further in the corporate policy as an active endorsement of modal shift, it could ease the introduction of BEVs into fleets. Otherwise, depending on the purpose the vehicles are used for, those 48 % of all reported vehicles in the sample could not only be substituted by BEVs, but also by other transport modes.

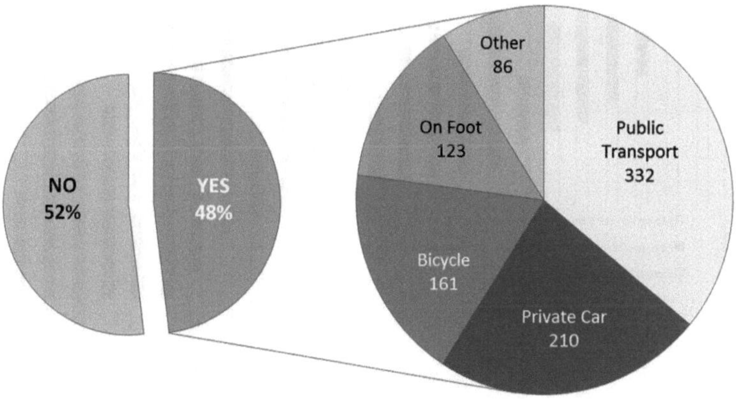


**Fig. 4** Applied modes of transport

One preferred scenario for the introduction of BEVs in commercial fleets is that on those rare occasions where the range requirement is above the maximum range of a BEV (in this example 140 km) the employee could use a different mode of transport: for long distances this would either be public transport, like trains, or conventional passenger cars. These vehicles could either be made available through commercial arrangements like car-sharing and corporate rental contracts or could include the employee's own vehicle as it is already practiced in 26 % of the complete sample.

## 7 Deployment Purposes

Another aspect that should be taken into account are the deployment purposes of the vehicles, which influence the mobility pattern of tomorrow. 47 % of the responding companies (only n = 750 for this question) reported that their vehicles were used for passenger transport only. This includes both transports of people as a service, like taxis, as well as company staff using the vehicles to visit clients. 43 % of the vehicles were deployed in mixed transport, meaning both goods as well as people were transported. This also included e.g. a service mechanic driving to a client with a spare part. The remaining 10 % of the vehicles in the sample were used for the pick-up or delivery of goods and material.

These three characteristics of vehicle deployment were mapped along the 27 industries (Fig. 5). Most of the industries seemed to have multi-purpose uses of their vehicles. Only two industries, namely passenger transport and nursing, used their vehicles solely for passenger transport. Only delivery services reported freight transport as their sole deployment purpose.

These results indicate two important determining factors influencing the substitution potential of conventional vehicles by BEV: on the one hand, the high

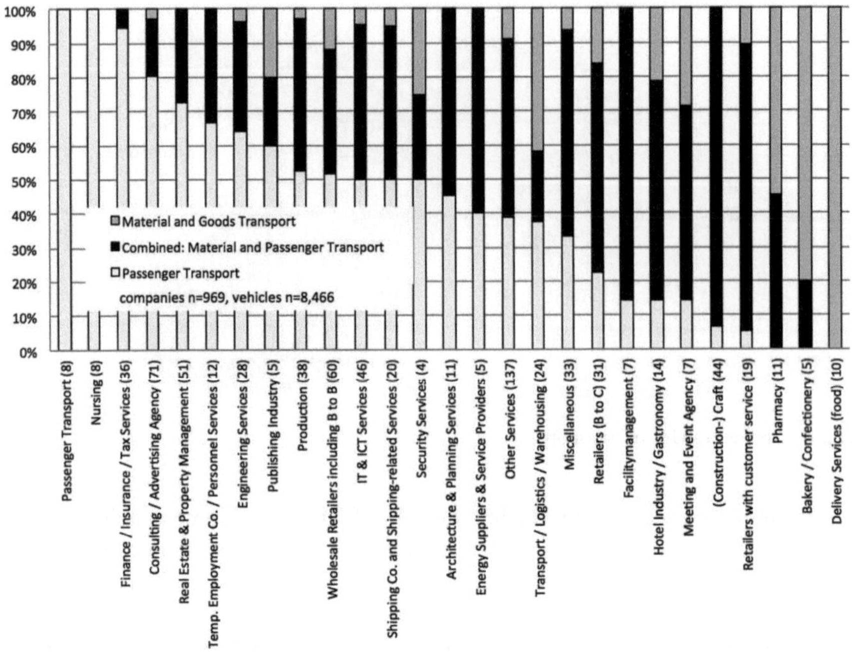


**Fig. 5** Deployment purposes of trips in the responding companies grouped by industries

potential for a modal shift in passenger transport from the individual car to other means of transport (e.g. public transport, bicycle) can lower the general demand for company-owned vehicles. On the other hand, there could be a lot of organisational barriers in the companies for such a change. These companies need further analysis for clearer evidence.

# 8 Procurement Strategies

Since the replacement cycles of vehicles are often determined by the ownership of the vehicle the survey also asked for the preferred strategy of procurement. Figure 6 provides an overview of the procurement strategies grouped by industries, differentiated into rental, purchase and lease.

There was no pattern to be found in the dependency of procurement and industries: there were ten different industries amongst those companies that lease 50 % or more of their vehicles. Informal information gathered from companies during the survey[1] indicated that purchase is preferred in those circumstances where

[1]A number of companies participating in the survey used the hotline provided by the University to clarify questions concerning the web-based survey.

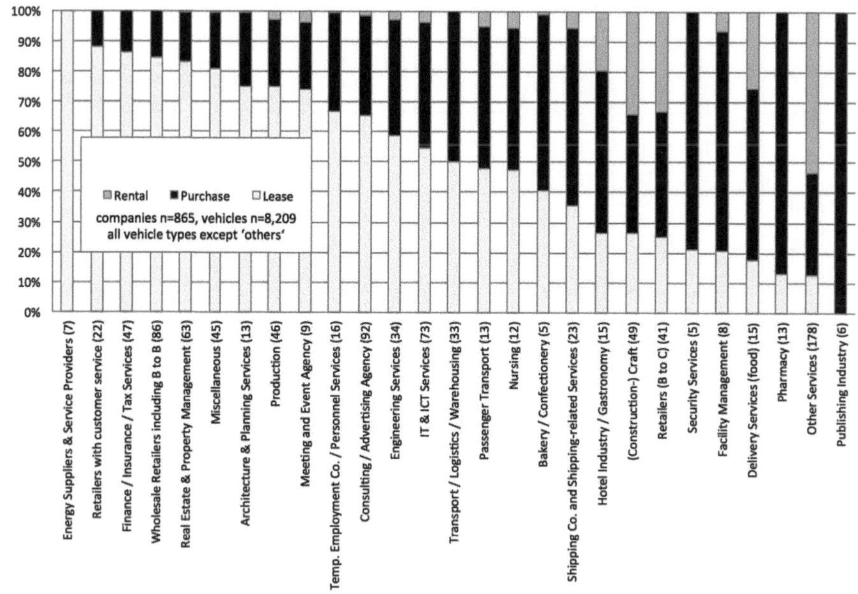


**Fig. 6** Deployment of procurement strategies in the responding companies grouped by industries

the use of the vehicle is likely to damage the car—like gardening or transporting certain goods—which would cause high penalties in a leasing contract. However, this could not be confirmed in the data for vehicles below 3.5 tonnes where retailers, energy services and manufacturing companies are well represented amongst those that primarily lease their vehicles.

## 9 Vehicle Replacement Cycles

Structural parameters like the vehicle replacement cycles influence the numbers of conventional vehicles which can be substituted by BEV over time. Figure 7 shows the time in months after which companies reported to replace a vehicle with a newer one.

843 participating companies reported how long they kept their vehicles in service. 33 % of the respondents keep their vehicles for 36 months, 28 % for 48 months and 16 % for 60 months. Only 2 % replace their vehicles after 12 months and 5 % after 24 months.

The results show that in any case the penetration of commercial fleets with BEVs, as desired by current government policy (NPE 2012), would take a considerable amount of time due to vehicle replacement cycles. Applied to the sample of this study, 2 % of the vehicles below 3.5 tonnes are due for replacement in the next 12 months, 5 % in the following year, and 33 % in the third year counted from

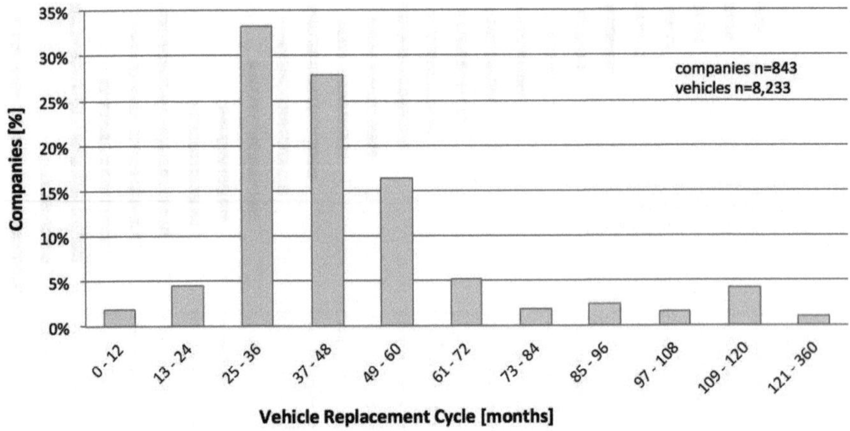


**Fig. 7** Deployment of vehicle replacement cycles in the responding companies given as intervals

today. In total, a maximum of 40 % of all reported vehicles in the survey will be replaced within the coming 36 months.

It follows, that in the next 3 years a maximum of only 40 % of all reported vehicles could be substituted by BEVs, assuming the necessary changes in organisational parameters, i.e. modal change or changes in vehicle fleet management.

## 10 Corporate Fleet Policy and Private Mobility

To evaluate the impact of range as a limiting factor for the use of BEVs, it is not sufficient to only assess the commercial use of a vehicle. Moreover, any additional use of the vehicles has to be considered. The use of vehicles may vary in companies, ranging from strict business-only use to private-only use or a combination of both. Business-only use covers journeys forming part of an employee's employment duties (such as journeys between appointments by a service mechanic) or journeys related to an employee's attendance at a temporary workplace. Everything else counts as private use. Combinations can be found where the car is taken home by the employee.

21 % of the companies in the survey only allowed the use of vehicles for business-related matters, in 79 % of the companies employees were allowed to use (some) vehicles for private use as well (Fig. 8). Only companies with very large fleets seem to be more restrictive in the use of their vehicles, allowing only business-related use. Apart from that no clear dependency of vehicle use policy on fleet size can be noticed.

One reason for private use of corporate vehicles is that they are often given to employees as a bonus to their salary. Kasten and Zimmer (2011) found out that

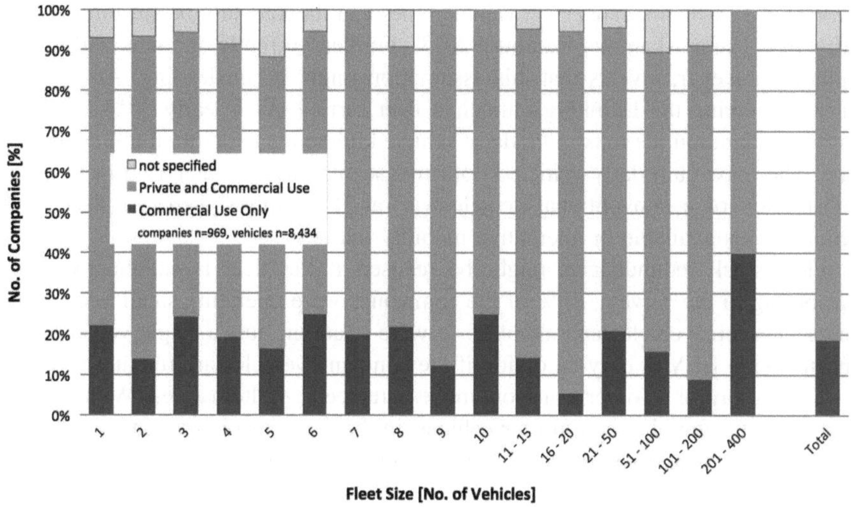


**Fig. 8** Distribution of vehicle use based on fleet policies as a function of fleet size

employees expected a fully usable car with a long range since they wanted to use them for a variety of different, unplanned activities and did not want to be restricted by the limited range of a BEV. Permitted private use could thus have a negative impact on the potential of BEV in an organisation.

## 11 Conclusions

The aim of this study was to analyse how organisational and structural parameters of commercial fleets influence the substitution potential of battery electric vehicles for conventionally-powered ones. In this context, the study examined what types of vehicles the fleets consists of and how these vehicles were being used and procured in the City of Hamburg. The response rate of almost 4 % was good for this kind of survey. After subtracting replies from those companies that did not run their own vehicles, the survey contained 969 valid questionnaires and information on 8,466 vehicles. 13 of 21 economic sectors and 27 of 88 industries are covered.

The majority of these industries preferred smaller cars, with seven industries also having a share of more than 50 % of larger vehicles. Still, all these vehicles, which made up 90 % of the sample, were below 3.5 tonnes gross vehicle mass and are currently available on the market in some form of BEV-equivalent.

Concerns about the range of these cars, as often expressed by users (Kasten and Zimmer 2011), seemed unfounded for most of the companies: within the sample analysed, 74 % of all passenger and light duty vehicles reported were used for less than 140 km a day. Studies suggest that vehicles fitting into this category are

"EV-ready", which means, such vehicle types can theoretically be substituted by a BEV (e.g. Handelskammer Hamburg 2013). Further, currently ongoing work by the authors, however, reveals that this assumption might be misleading. An in-depth survey covering the Hamburg municipal fleet carried out in early 2014 also took into consideration the maximum daily vehicle kilometres in addition to the average one. The maximum daily vehicle kilometres seems to be the main obstacle to the substitution of a conventional vehicle with a BEV, unless there are fleet stock redundancies available or alternative mobility options.

Fleet stock redundancies might be realised if the fleet is sufficiently large. According to the survey, 70 % of the companies were operating small fleets of no more than four vehicles and about 25 % were operating just a single vehicle. This means that a BEV is only an option if the company vehicles are running less than 140 km a day or if other transport modes are seen as alternatives. With slightly more but still few, i.e. two to four vehicles in the fleet, keeping a conventional car for the rare occasions of long haul travel might guarantee a high level of redundancy.

Alternative mobility options like other modes of transport, on the other hand, have the potential to foster the introduction of BEV in company fleets by securing a fallback-option for the occasions where a driving range above 140 km a day is required. 48 % of the responding companies already used additional modes of transport and could benefit from this flexibility in mobility patterns in their transport policy strategy.

This is especially true when taking into consideration the predominant use of the vehicles: 47 % of the respondents used their vehicles primarily for passenger transport, which is usually not restricted in the use of other modes of transport. However, it is also possible that the companies may decide not to purchase a company vehicle at all. For example, inter-corporate car-pooling can be implemented where different companies share one car-pool with different options of vehicle types and engines to perform also longer trips (e.g. pilot project car2share, Hamburg). For companies in rural areas where car-sharing has not been established yet and public transport is scarce plug-in hybrid electric vehicles (PHEV) might be a reasonable option as long as the performance of BEV is as it currently is.

Besides the technical aspects of vehicle substitution this study also looked at the temporal aspect of a possible penetration of commercial fleets by BEVs. The procurement process of companies is determined, among others, by the vehicle replacement cycle, either ruled by leasing contracts or other economic reasons. At best, and in case all organisational parameters are realised, a maximum of 40 % of all reported vehicles in this sample could be substituted within the next 3 years.

An additional aspect that needs to be considered in the assessment of the substitution potential is whether the vehicles in the respective fleet are used solely for conducting business or if employees are allowed to take the cars home for private use. Often company cars serve as a bonus system for employees, where employees are allowed to use the car outside working hours, sometimes also combined with further incentives like unlimited mileage and fuel. Kasten and Zimmer (2011) show that this policy often leads to an intensive use of the vehicles outside business hours including long distances covered, e.g. at the weekend. Nearly 80 % of the companies in the sample used this corporate policy which is not conducive to the market entry of battery electric cars into commercial fleets.

To put it in a nutshell, the survey showed that an isolated analysis of technological developments in the field of battery electric vehicles falls short in the prediction of the substitution potential in commercial fleets. While organisational parameters have an impact on the technical feasibility of substitution, structural parameters, like procurement strategies or the rate of renewal of vehicles in the companies, influence the period in which the substitution potential could be realised.

With numerous new business models in the mobility sector, like car-sharing or internet-based mobility services, a modal change could increase the potential to substitute conventional vehicles in companies with BEVs. At the same time, economic reasons, caused by increasing fuel prices or a (mostly) customer-induced demand for 'greener' transport, could change the current mobility patterns. The results showed that for 50 % of the companies analysed a modal shift would require a profound change in their organisational transport practice.

However, it yet lacks clear evidence that such efforts, e.g. the separation of passenger and goods transport, could be a real step forward, either environmentally or economically. Further research on this question is recommended, especially as the survey showed a huge variation of factors and their levels, like the deployment purposes of the vehicles or the vehicle types represented in the fleet, which differ between industries. Gaining a better understanding of the composition of fleets in a wide range of commercial sectors, as well as of vehicle use and procurement strategies, would enable a more precise forecast of the potential for BEV-penetration in commercial vehicle fleets. This may also help car manufactures to better estimate the future demand for different BEV types. Also, recommendations for the development of recharging infrastructure and policy measures could become more accurate.

**Acknowledgments** This work was gratefully funded by the German Federal Ministry of Transport and Digital Infrastructure.

## References

Bundesministerium für Verkehr Bau und Stadtentwicklung (BMVBS) (2011) [now: German Federal Ministry of Transport and Digital Infrastructure]: Was erwarten künftige Nutzer von elektrischen Fahrzeugen? [What do future users expect from electric vehicles?]. Report from the NOW working group on social sciences. Berlin

Bundesministerium für Verkehr Bau und Stadtentwicklung (BMVBS) (2012) [now: German Federal Ministry of Transport and Digital Infrastructure]: Kraftfahrzeugverkehr in Deutschland 2010 Schlussbericht. Braunschweig

Federal Republic of Germany (2009) German Federal Government's National Electromobility Development Plan. Berlin

Gnann T, Plötz P, Zischler F, Wietschel M (2012) Elektromobilität im Personenwirtschaf tsverkehr – eine Potenzialanalyse [Electric Mobility in Commercial Passenger Transport - a potential analysis]. Working Paper Sustainability and Innovation No. S 7/2012. Fraunhofer ISI. Karlsruhe

Handelskammer Hamburg (2013) Einsatzpotenziale für Elektrofahrzeuge in der Hamburger Wirtschaft. Hamburg

Holzer H (2014) Warum sich Elektroautos noch immer nicht rechnen [Why electric vehicles still are not cost-effective]. In: Welt D (ed) online access on 2014-08-14 17:12. http://www.welt.de/125305573

Kasten P, Zimmer W (2011) $CO_2$-Minderungspotenziale durch den Einsatz von elektrischen Fahrzeugen in Dienstwagenflotten [carbon dioxide reduction potential in commercial fleets from the use of electric vehicles]. Ergebnisbericht im Rahmen des Projektes "Future Fleet" AP 2.7. Öko-Institut, Berlin

Nationale Plattform Elektromobilität (NPE) (2012) Fortschrittsbericht der Nationalen Plattform Elektromobilität (Dritter Bericht) [Third status report of the national working group for electric mobility]. Gemeinsame Geschäftsstelle Elektromobilität der Bundesregierung (GGEMO). Berlin

Trümper SC (2013) Battery-electric demonstrations in early-adopter-fleets. In: Proceedings of the EVS27 international battery, hybrid and fuel cell electric vehicle symposium, Barcelona, Spain, November 17–20

# Who Wants to Be a Millionaire? The Lead Market Potential of Electric Mobility

Noreen E. Werner

**Abstract** This paper examines the conditions of global dominant innovation design considering e-mobility and the emergence of the lead market in different countries. The development of lead markets is linked to multiple, country-specific factors, which influence the lead market potential and position through the decreasing of national differences over time by benefits of the best compromise of an internationally similar innovation design. Due an early market entry, firms can reach first mover advantages, which they have to protect against followers. Especially eco-innovations could reach a state of a global dominant design, because they are developed to tackle the climate change. The development of globally sold e-mobility products, e.g. the electric motor, the battery or the whole vehicle is more likely. Therefore, the strategy to strengthen a lead market for e-mobility seems to be promising. With the help of a principal component analysis (PCA), different country-specific indicators were applied in the model of five lead market factors. But lead markets change quickly and the given theoretical concept is only estimating opportunities and not forecasting technology developments. After the evaluation the USA obtain the 1st lead market position according to e-mobility, followed by Germany and France. The advantage of the USA is mainly a result of demand incentives and a large sales market. In Germany supply side indicators are more pronounced. The position of France mainly results from well advanced supply and demand side measures in average.

**Keywords** Lead market · Dominant innovation design · Innovation · International diffusion · Principal component analysis

N.E. Werner (✉)
Centre for Regional and Innovation Economics, University of Bremen, Wilhelm-Herbst-Str. 12, 28359 Bremen, Germany
e-mail: n.werner@uni-bremen.de



D. Fornahl and M. Hülsmann (eds.), *Markets and Policy Measures in the Evolution of Electric Mobility*, Lecture Notes in Mobility,
DOI 10.1007/978-3-319-24229-3_12

# 1 Introduction

The automotive industry is moving towards a transition point from conventional to electric vehicles. In Germany, the export of vehicles and related components is one of the largest industries with a high number of qualified workforces. To obtain a leading position, the market share of electric cars and according processes from German Original Equipment Manufacturer (OEM) has to increase because the goal of the German government is to become the lead market for e-mobility (BMVI 2011 amongst others). Other companies enter the automobile market with local joint venture partners; they are subjected to increasing pressure from the aggressive pricing strategies.

Lead markets are local markets, which use a specific innovation design earlier than other countries. Due to an early market entry the companies could achieve technology-specific experiences earlier, which lead to competitive advantages. Lead markets are not necessarily the markets where the innovation was invented. The lead market approach considers local market conditions (e.g. the ability to export and company competitions), which explain the commercialization of innovations more than the technology gap (Beise 2006). For example, the antilock brake system (ABS) was invented in Great Britain in the 1930s for airplanes to prevent the wheels from brake-locking on wet surfaces. Later on, the technology was used in vehicles and was finally adopted and developed to a global dominant design by German industries (Beise 2006).

Eco-innovations like fuel-efficient and electric cars are supposed to have large international market potential because of global requirements to tackle climate change effectively. For economic reasons, it cannot be expected that in case of e-mobility consumer-specific solutions (e.g. several motor and battery technologies) diffuse. Therefore the chance that other countries follow a national innovation design accelerates with global needs and reducing costs. Even the transfer of political regulations can have lead market potential, especially for eco-innovations (Cleff and Rennings 2011; Quitzow 2013).

The hypothesis is that a dominant innovation design could lead the way of firm's or even nation's success to reach the goal of one million registered electric vehicles earlier than other countries. The paper investigates national conditions considering e-mobility and the emergence of the lead market. After theoretical considerations, we will test the lead market potential of different countries with the help of a principal component analysis (PCA) and Statistical Package for the Social Sciences (SPSS). We conducted and analyzed metric data to have a close look at country-specific advantages and disadvantages of several country markets and their challenges for e-mobility.

## 2 Lead Market Strategy

To determine the lead market potential for electric vehicles, it is presumed that the competitive position of countries for a certain innovation design can be derived from country-specific market conditions. Patterns of adoption (individual decision-making process) and diffusion (cumulated adoption decisions) can be traced back to country-specific phenomena, such as level of income, market formalities, technological know-how or demand (Rogers 2003; Ganesh and Kumar 1996). The more recent developed technologies or processes meet the requirements of a broad range of customers, the faster they will be able to diffuse. The task is to develop globally introduced innovations, which offer competitive advantages through standardizations by meeting the needs for responsiveness and integration in firm networks and platform strategies to reduce total costs (Beise 2001).

Prior studies showed that dominant innovation designs most likely emerge from one or more radical innovations. Radical innovations ran through a selection process of trial-and-error and have been evaluated as the best technological compromise (for an overview see Murmann and Frenken 2006). Due to learning and scale effects in production as well as network externalities, the innovation design can achieve higher market shares and evolves into a dominant design (Abernathy 1978; Rosenbloom and Christensen 1995, amongst others). Along with the diffusion and the successive elimination of competitors on national and international markets, it is referred to as a global dominant innovation design. In conclusion, "[…] lead markets are countries that first adopt a globally dominant innovation design; they lead the international diffusion of an innovation and set the global standard" (Rennings and Beise 2003, p. 5). If a country becomes a lead market, depends on market conditions and country-specific factors.

Thus, lead markets do not necessarily emerge in those countries, in which the innovation design was originally developed. It is occasionally sufficient to pick up an innovation and "cultivate" it in a way that the innovation design can be successfully brought to the market (Beise 2006, p. 138). A possible explanation for this might be that while countries prefer specific fields of technology and thus have a different technological expertise, the generation of knowledge does not guarantee an innovation's market success automatically (Golder and Tellis 1996). Companies can also appear as "second" innovators at the beginning of the adoption process, if they are capable of gaining knowledge and learning from it (learning by doing and learning by using) (Rosenberg 1982; Beise 2006). The case of Nokia shows that even firms, which are not associated with a specific technology, can become successful market actors. The company of Nokia started as a paper manufacturing plant and shifted its business activities to the production of rubber boots and automobile tires. With the emergence of mobile telephony in the 1980s, Nokia stepped into the market and became an international successful pioneer. Nokia was able to sustain its position over a long time (Nokia 2014; Beise 2006).

## 2.1 Types of Country Markets

Taking into account the prior considerations, countries can be arranged in a matrix of four market types (Fig. 1). As a complement to the performance of country markets further attributes can be determined by using the concept of first mover advantages by Lieberman and Montgomery (1987, 1998) and the management of multinational firms by Bartlett and Goshal (1990). The first-mover-advantage-approach considers the optimal point of market entry in dependence on the pace of technology and market evolution (Suarez and Lanzolla 2005). From a managerial perspective, the strategic decisions of subsidiaries concerning local innovations and local resources and capabilities are significant for multinational companies (MNC) (Bartlett and Goshal 1990).

Lead markets (upper right) are countries, which have an innovation design with clearly identifiable, national characteristics. The importance of local innovations on such national markets is very high. The domestic market has a comparatively high lead market potential, if these innovations can be integrated into international innovation and marketing strategies easily (Beise 2006). Lead markets exist in different countries for different product categories. Furthermore, an industrial sector

**Fig. 1** The different types of country markets and their underlying theory (own figure reproduced from Beise 2006; Suarez and Lanzolla 2005; Bartlett and Goshal 1990; Lieberman and Montgomery 1987)

can have several lead markets for different segments. For example in the automobile industry, the leader for premium vehicles is Germany, for Sport Utility Vehicles (SUV) it is the USA and for luxury sport cars it is Italy (Johannson 2009). They are characterized by a high pace of market and technology evolution due to "[…] natural resource endowments, technological know-how, and labor skills" (Johansson 2009, p. 18). In several aspects the evolution on lead markets is very fast; therefore the situation for companies is described as "rough water". The high levels of demand and competition cause firms to implement global strategies (leader), as being the first mover. Creating demand requires major investments before a critical mass of adopters is reached, so that the adoption rate could increase independently (Rogers 2003).

The so-called lag markets (lower right) are follower markets, which adopt firstly the dominant innovation design of the lead market. Lag market are as innovative and technologically advanced as lead markets, but their domestic innovations often do not succeed on an international level. The international relevance of market-based innovation designs and the slow pace of technology evolution on lag markets do not match with international marketing strategies of MNC. Therefore, on lag markets the "market leads". MNCs tend to focus on export-based and decentralized marketing strategies (contributor) instead of using domestic resources and skills for innovation development. However, lag markets are able to become great competitors of lead markets. They have appropriate competencies and can closely observe the lead market, which impacts the speed of the learning curve for the adoption of an innovation design and reduces the uncertainties of an innovation (Meyer-Krahmer and Reger 1997).

Large, isolated markets (upper left) prefer mainly local innovations. However, they cannot be easily integrated into international strategies due to a high amount of innovation-specific characteristics. The high importance of local innovations increases a technology leading development along the home market costumer preferences. The advantage of domestic firms results from a great sales market, which is often protected against foreign competition through official regulatory procedures like in USA or China (Beise 2006). Along with large, isolated costumer markets and potential official regulatory procedures the danger of high investments and extensive coordination among the subsidiaries could emerge. Acting in such kind of market implies operating in a "black hole" (Bartlett and Goshal 1990).

The fourth type of country markets are peripheral markets (lower left). Typically, these markets are dominated by domestic firms due to their know-how of national preferences and long-term experiences with costumer needs (implementer). Such markets normally have a small costumer market, where mainly nation-specific strategies could effectively applied (Beise 2006). As a result of low competition the pace of technology evolution is slow, which characterizes acting in "calm water". The Integration of innovations in international strategies is not suitable, mostly due to a low level of local resources and capabilities. Additionally the need of innovations is very low, what protect the leadership of domestic firms longtime.

## 2.2 Lead Market Factors

The question is, why do certain innovation designs diffuse on an international level, despite national differences? According to the research of Beise (2001, 2006), Beise et al. (2002), Rennings and Beise (2003), Beise and Cleff (2004) amongst others, multiple country-specific factors ("lead-market-factors") can be considered, which cause the decrease of national differences over time through benefits of an internationally similar innovation design. The lead-market-factors influence the international competitiveness of innovations and indicate the national leading capability of country markets.[1] Due to homogeneous governmental, social, economic and cultural influences, the diffusion into other countries is more likely, the higher the similarities of both markets. In the following section the lead market factors are briefly described.[2]

- **Demand Advantage**: The anticipation of future global trends is the emphasis of a lead market. By successfully predicting certain developments in global market needs, innovation designs can be aligned to meet these requirements in early stages of the product development and lead to first mover advantages. Therefore, the demand advantage relates to an early adoption of an innovation design, which becomes a global trend. "[…] The difference between different countries' markets does not lie in the direction, in which they develop, but merely in the speed with which they move in the direction of the global trend" (Cleff and Rennings 2013, p. 8).
- **Price Advantage**: This factor accounts for the decrease of the relative price of a preferred national innovation design in comparison to foreign designs. In competitive respectively lead markets the price competition arrives with the innovativeness and effectiveness dynamics of companies. Moreover it could be argued that the main reason for variety in using specific technologies goes back to price differences (Beise 2006). Price advantages only occur, if price differences between country markets are "large enough to exceed the transaction costs incurred in changing over to the cheaper innovation design" (Cleff and Rennings 2013). Another possibility to achieve a price advantage is the anticipation of international factor price changes.
- **Transfer Advantage**: Transfer advantages can emerge by demonstrations of the usefulness and reliability of an innovation design combined with an intense transaction and communication between the lead market and foreign markets. This can be explained by the diffusion of information and the successful decrease of uncertainties regarding the reliability of the product in foreign

[1] Newer studies assume additionally the existence of a sixth factor (regulation advantage) (Köhler et al. 2014), which will not be considered in this article. Further, there exist external impacts through research and development (R&D) infrastructure and coincidence (Beise 2006).

[2] For a detailed description and a list of according variables see Beise (2006).

markets. International reputation of firms or the communication infrastructure often is used to describe transfer advantages.

- **Export Advantage**: The export advantage describes the potential of firms to create products for domestic and foreign markets by inclusion of foreign preferences in nationally preferred innovation designs (Beise 2006). As long as foreign users do not miss a significant function or utility, the barrier of changing to another innovation design is lowered. Therefore, companies tend to standardize products within their global network.
- **Market Structure Advantage**: The market structure advantage describes a high competitive pressure on the domestic market, which forces firms to improve their innovation designs in order to be able to compete. As the original innovation is improved several times, successful innovation designs tend to fit the needs of the customer on national as well as on international level. In lead markets there is a strong firm competition, which forces the market and technology evolution. The lead-time advantages on their competitors (first mover advantage) is the most frequently used protection method.

## 3 Relevance of the Lead Market Analysis for Electric Mobility

Is the lead market concept really suitable for electric mobility? Yet, for most innovations in the automobile industry this was not the case. The demand for globally standardized automobiles or components is rather small, due to the diversity of various national markets. For instance, the development and production of automobiles is heavily dependent on particular official regulatory procedures, the conditions of the road infrastructures and varying customer preferences in terms of design and performance (Beise 2006). Therefore, only a few components are applicable in more than one country (e.g. ABS, airbags and direct diesel injection).

An exception may be the case of battery electric vehicles (BEV), which benefit from e.g. efficient battery technology, the deployment of charging infrastructure, the development of electronic systems and smart grids. The international standardization of these components would be advantageous as they could contribute to the evolution of a certain electric vehicle into a dominant design. Climate change forces global needs like the fuel efficient vehicles as well as the independence on oil. At the moment, BEVs have no country-specific differences in design, equipment and performance, but they have national characteristics. For example, the design and performance of Nissan Leaf or Toyota Prius are very different to traditional demanded car segments in the USA (SUVs, Pic ups), but nevertheless they are very successful on the U.S. market. This supports the assumption that electric vehicles may overcome the national design and performance preferences of conventional cars.

# 4 Methodology

According to the lead market approach, the basic dimensions (lead market factors), which characterize the country-specific market conditions in relation to the innovation design are already theoretically embedded. Regarding to this, 51 variables matching the innovation project "electric mobility" were collected and analysed. Only metric data from secondary statistics, like international databases or the national statistical bureaus of the respective countries (e.g. OECD, Worldbank, Eurostat and UN) were utilized. The included 29 OECD countries were chosen in consideration of data availability and a similar technological and social stage of development, regarding the automobile industry. The OECD countries Israel, Iceland, Ireland, Luxembourg, Chile and New Zealand are excluded.

The prognosis model is based on an analysis of the multi-causal inter-relationship with non-weighted lead market factors. On this basis, the principal component analysis (PCA) was applied. The overall goal of a PCA is the reduction of a large amount of variables to a smaller set of common dimensions through the building of groups of highly correlated variables as well as the validation of the model (Backhaus et al. 2006). The PCA requires individual decisions by the researcher regarding the selection of the metric data and the determining of components. The components are weighted during the evaluation process in SPSS automatically. Following steps for the PCA in SPSS are necessary.

(1) Based on the theoretical assumptions, variables are compiled, which correspond to the investigation object of e-mobility. The raw data matrix' variables are based on different measuring units. In order to create a comparable data basis, the standardization of the variables (called Z-transformation) is mandatory.
(2) The basic information for the PCA is the correlation matrix with the Pearson product-moment correlation coefficient, giving a value between +1 and −1, where 1 is a total positive correlation, 0 is no correlation, and −1 is a total negative correlation. Those variables that share similar dimensions are highly correlated in this model. Variables with lower correlations than 0.6 will be deleted.
(3) The PCA requires a Kaiser-Meyer-Olkin (KMO) measure of sampling adequacy higher than 0.5 for the variable set. The KMO-criteria determines whether the data set is suitable for the PCA.
(4) To discover and summarize the pattern of inter-correlation among the variables, the scree plot shows the break point in a graph of the eigenvalues at which the eigenvalue is plotted against the number of principle components (Backhaus et al. 2006).
(5) The communalities represent the proportion of the variance in the original variables that is accounted for by the factor solution. The communality for each variable should be higher than 0.5 (Backhaus et al. 2006). The task is repeated until the components contain the most suitable communalities.
(6) The oblimin rotation allows any correlation between the factors with values higher than 1, because the factors are influenced by each other.

(7) For measuring the factor values a multiple regression method is conducted, which leads to standardized values with an arithmetic mean of 0 and a standard deviation of 1 (Backhaus et al. 2006). The regression coefficient is calculated as the product of the weighting factors in the coefficient matrix and the standardized variable values of the raw data. The higher the regression coefficient the higher is the lead market potential. Negative values describe an underperformance in comparison to all other countries. A value of 0 stated an average performance of the factor and positive values represent an above-average performance of the country factor.

# 5 Results

The PCA groups the information of 18 variables in five principle components. All other variables have to be removed because of communalities under 0.5. The KMO measure of sampling adequacy is 0.64, which is acceptable, but not optimal. The scree plot shows that choosing five components is acceptable, because the curve flattened down after the fifth factor. The five components signify 95 % of the total variance of the data set. Table 1 represents the output of five lead market factors and their describing variables. The following Table 2 shows the countries and relating regression coefficients, which located the countries in order to their lead market potential. Further it shows the skewness of distribution, which measures the strength of the factors and how the distribution is pulled to one or the other side. To determine the lead market position, the price advantage is the strongest factor followed by the demand advantage. The skew values indicate strong differences between the countries especially in price, market structure and demand. None of the tested countries has more than one factor advantage. However, the lead market theory insists that there is no mono-causal explanation because the countries have specific innovation design requirements and they prefer products because of economic factors (Beise 2006). If economic factors like income, relative use or relative prices differ extremely, the countries will prefer different innovation designs. But a lead market only exists, if these differences will be compensated for. Therefore, the performance of the innovation design is not relevant, but the country-specific economic structure.

## 5.1 *Price Advantage*

The strongest component in the examination is the price advantage. The variables are highly positive correlated. A positive indicator for a country means that the country-specific price level is above the average price level of other countries and vice versa (Cleff and Rennings 2013). In the study the price advantage is

**Table 1** The output and allocation of variables to the lead market factors (OECD 2013, Worldbank 2012, 2013; UNcomtrade 2013; OICA 2013; ACEA 2013; AFDC 2013; IRI 2013; Statista 2013; REN21 2012, own calculation)

| Variables | Principle components | | | | |
|---|---|---|---|---|---|
| | Price advantage | Transfer advantage | Market structure advantage | Export advantage | Demand advantage |
| Indices of end-user real energy prices for households compared to 2005 (total energy) | −0.983 | | | | |
| Indices of end-user real energy prices for industry compared to 2005 (total energy) | −0.979 | | | | |
| Export volume of high-tech products | 0.807 | | | | |
| Installed capacity for electricity generation from renewables in mega watts | 0.756 | | | | |
| Volume of vehicle production by country | 0.723 | | | | |
| Volume of new vehicle registrations by country | 0.638 | | | | |
| Purchasing power parities for GDP and related indicators: comparative price levels | | 0.946 | | | |
| Income per capita | | 0.925 | | | |
| Patent volume technologies specific to propulsion using electric motor (e.g. electric vehicle, hybrid vehicle) | | | 1.020 | | |
| Patent volume technologies specific to propulsion using internal combustion engine (ICE) (e.g. conventional petrol/diesel vehicle, hybrid vehicle with ICE) | | | 0.960 | | |
| Number of world companies in automobile industry by country | | | 0.801 | | |
| Vehicle export units per 1,000 capita | | | | 0.965 | |
| Vehicle import units per 1,000 capita | | | | 0.785 | |
| Household expenditure for vehicles | | | | | 1.034 |
| Number of charging stations | | | | | 0.920 |
| Roads total network in km | | | | | 0.815 |
| Inward FDI stocks | | | | | 0.807 |
| Sales volume of battery electric vehicles | | | | | 0.774 |

Extraction method: principle component analysis; Rotation method: Oblimin with Kaiser-normalization

characterized by indices of end-user real energy prices for industry compared to 2005[3] (total energy), indices of end-user real energy prices for households compared to 2005 (total energy), export volume of high-tech products, installed

[3] 2005 = 100.

**Table 2** Lead market positions per factor by regression coefficient per country (own calculation)

| Country | Price advantage | Transfer advantage | Market structure advantage | Export advantage | Demand advantage |
|---|---|---|---|---|---|
| Australia | −0.215 | **1.142** | −0.433 | −0.539 | −0.226 |
| Austria | −0.268 | 0.676 | −0.295 | 0.107 | −0.181 |
| Belgium | **0.223** | 0.893 | −0.586 | **3.776** | −0.069 |
| Canada | 0.111 | 0.667 | −0.427 | 0.313 | 0.006 |
| China | **4.952** | −2.543 | 0.090 | −0.891 | **0.538** |
| Czech Republic | −0.400 | −0.985 | −0.215 | 0.691 | −0.414 |
| Denmark | −0.020 | 1.096 | −0.440 | −0.345 | −0.489 |
| Estonia | −0.534 | −0.711 | −0.394 | −0.543 | −0.489 |
| Finland | −0.381 | 0.590 | −0.331 | −0.561 | −0.459 |
| France | 0.021 | 0.292 | 0.298 | −0.193 | 0.369 |
| Germany | 0.119 | −0.051 | **1.922** | 1.083 | **0.719** |
| Greece | −0.568 | −0.254 | −0.418 | −0.761 | −0.417 |
| Hungary | −0.607 | −1.054 | −0.316 | −0.313 | −0.348 |
| Italy | −0.280 | 0.064 | −0.274 | −0.496 | −0.172 |
| Japan | 0.029 | 0.260 | **4.466** | −0.617 | 0.019 |
| Mexico | −0.147 | −1.046 | −0.409 | −0.617 | −0.284 |
| Netherlands | 0.132 | 0.722 | −0.401 | 0.253 | −0.086 |
| Norway | 0.189 | **2.047** | −0.503 | −0.514 | −0.294 |
| Poland | −0.582 | −0.750 | −0.323 | −0.640 | −0.349 |
| Portugal | −0.442 | −0.441 | −0.384 | −0.428 | −0.365 |
| Slovak Republic | −0.210 | −0.742 | −0.337 | **2.074** | −0.491 |
| Slovenia | −0.522 | −0.514 | −0.324 | **1.392** | −0.419 |
| Spain | −0.386 | −0.321 | −0.270 | −0.051 | −0.109 |
| South Korea | −0.064 | −0.689 | 0.387 | −0.239 | −0.482 |
| Sweden | −0.228 | 0.819 | −0.146 | −0.223 | −0.291 |
| Switzerland | 0.030 | **1.619** | −0.363 | −0.145 | −0.214 |
| Turkey | −0.477 | −1.605 | −0.358 | −0.831 | −0.447 |
| United Kingdom | −0.197 | 0.343 | −0.221 | −0.068 | 0.536 |
| USA | **0.725** | 0.476 | **1.003** | −0.678 | **4.907** |
| **Skew** | **4.621** | −0.**283** | **3.684** | **2.433** | **4.517** |

capacity for electricity generation from renewables in megawatts, the volume of vehicle production by country and the volume of new vehicle registrations by country.[4]

[4]Considered are only first-year admissions with a mileage of 0 km.

For the production of alternative engines and the demand of the products the energy prices play a key role. Electric cars only have a price advantage, if the energy price is lower than the gasoline price. With increasing energy prices the export of high-tech products e.g. electric, safety systems and battery technology increases as well as the vehicle production and registrations. Companies and households search for alternatives for conventional energy production technologies, e.g. by using solar and wind energy. Likewise, the installed capacity of renewable energies is positively correlated with increasing energy prices. It also has positive influences on the volume of vehicle production and registration because the demand reacts positively, with an above average for new and efficient vehicles. Possibly, there has to be considered governmental legislations and subsidies, which accelerate firms to develop energy-efficient innovations.

The regression factor value shows that China owns the highest price advantage followed by the USA and Belgium. The energy price index of the USA (101.3) and Belgium (105.3) is below the average of 122.5 (compared to 2005) of all tested countries (OECD 2013). For China there are no values available, but the energy price for industrial consumers nearly is double that of the USA level (IEA 2013). The predominant energy-intensive industries and low energy prices in the USA do not enhance energy efficiency. This is compensated by high absolute values in energy capacity of renewable energies as well as in production and registration volumes of vehicles through great domestic consumer markets. In other countries (e.g. Germany, Japan) the energy costs for industry and households are higher, which leads to a disadvantageous ratio between costs and production. Belgium has a price advantage because of the traditionally strong automobile production industry in the EU and OEMs like Volkswagen/Audi, General Motors, Toyota and Ford (partly with European headquarters). These OEMs are mainly export oriented, which makes Zeebrugge to an important international hub for the overseas trade of motor vehicles and motor vehicle components (GTAI 2014). Additionally, Belgium has a high share of newly registered vehicles, which is surely a result of the high fuel price (1.70 €) and the increasing demand of new energy-efficient vehicles (Worldbank 2013).

## 5.2 *Transfer Advantage*

In this study, income per capita and comparative price level[5] load on one component, which was not expected. Measuring the transfer advantage is mainly a qualitative task (e.g. in case of measuring the reputation of firms), which limits the selection of metric data. Normally the income is applied to describe the demand behavior, whereas the comparative price level represents price and cost structures

[5]It measures the differences in price levels between the countries by computation the ratio of purchasing power parities for private final consumption expenditure to exchange rates (OECD 2013).

(Beise 2006). However, high-income countries are often trendsetters for new technologies and will be observed by other countries. Studies have shown that the intention to buy electric vehicles is higher in high-income and multi-person households with more than two vehicles than in other households (Peters and Dütschke 2014). The adoption rate could increase with high income, which accelerates the transfer of an innovation design. If the critical mass of adoption is reached, the price will decrease and the differences between the countries will be compensated. Therefore, the comparative price level signifies the ability of countries to produce and offer specific goods by lower opportunity costs. We argue, the higher the comparative price level, the higher the income per capita (and adoption rate) and the higher the transfer advantage.

The skewness of distribution shows that differences between the countries are rather low. For lead markets this is a necessary condition because similarities in income and comparative price level lead to faster adoption rates between the countries. In the model, Norway, Switzerland and Australia acquire the advantage of transfer, which means that they could have an impact on the adoption rate of other countries. The private consumption expenditures including indirect taxes as well as the income per capita are higher than in other countries. Norway is a good example for a country with demand incentives, which accelerate the adoption of BEVs despite a small automobile industry. Countries with a high level of economic development according to the user potential will be deemed to be reliable and qualitatively demanding. But the similar component values of the transfer factor push the presence of other competitive factors like price or export.

## *5.3 Market Structure Advantage*

At present, the market structure variables for e-mobility have only limited validity because the manufactures are often start-ups, metric data are not available and publication and patent data for e-mobility are not long standing and limited in their validity to technological quality and change. The output of the PCA shows three relevant variables: patent volume technologies specific to propulsion using electric motor (e.g. electric vehicle, hybrid vehicle), patent volume technologies specific to propulsion using internal combustion engine (ICE) (e.g. conventional petrol/diesel vehicle, hybrid vehicle) and the number of world companies in automobile industry. Surely, patent applications are only one option for innovation protection, but mechanisms like secrecy or complex product designs are not measurable. It is argued that countries with relatively high patent volumes for BEV and ICE have an advantage through a long standing reputation and continued R&D activities in the automobile industry, which will be observed by other countries. The number of patents is positively correlated with the number of world companies in the automobile industry per country. As already mentioned, a high market competition is a key factor of lead markets because the costumers are more selective, the quality of innovations is higher and refined innovations that better fits costumers' preferences are more likely (Beise 2006). Additionally e-mobility is a cross-sectional technology and has applications in many areas and leads to productivity growth in all important sectors.

After calculation, Japan has the greatest market structure advantage followed by Germany and the USA. The countries are part of the world's leading automotive nations with the highest amount of OEMs and production rate. Japan applies a total of 467 technologies specific to propulsion using electric motor (e.g. electric vehicle, hybrid vehicle) for patent in 2010 (OECD 2013). Germany holds 167 and the USA achieved 79 patents (OECD 2013). In 2010, 697 patent applications for technologies specific to propulsion using internal combustion engine (ICE) (e.g. conventional petrol/diesel vehicle, hybrid vehicle with ICE) were applied in Japan, 505 patents in Germany and 247 patents in the USA (OECD 2013). Additionally, the production of lithium-ion batteries is mainly still in the hand of Japanese and South Korean manufactures. Japan's domestic market covers the whole supply-chain of e-mobility, while countries like France and the USA import most components (Roland Berger and fka 2014a). German OEMs practice mainly joint ventures to use the know-how of 1st tier suppliers. In Japan, there are situated 30 leading companies of the automobile industry out of 1,500 world's leading companies according to R&D activities (e.g. Toyota, Nissan, Mazda, Suzuki, Mitsubishi, Honda, Yamaha) (IRI 2013). In Germany, there are 14 (e.g. Volkswagen, BMW, Daimler, Porsche, ZF, Robert Bosch, Continental) and 20 companies are located in the USA (e.g. Ford, General Motors, Goodyear, Tesla Motors, Johnson Controls) (IRI 2013). The ratio stays almost the same by considering e-mobility related industries like electronic components, industrial engineering and chemicals.

## 5.4 Export Advantage

The export advantage is represented by the number of vehicle exports per 1,000 capita and the number of vehicle imports per 1,000 capita. Countries with high export rates are sensitive to foreign needs; often have long-term export experiences and innovation requirements that will be shaped by the global market (Ohmae 1995). The high import rate could represent a high propensity to consume and a competitive behavior of firms.

The export advantage is led by Belgium followed by the Slovak Republic and Slovenia. As already mentioned, Belgium has a high standing in the EU automobile industry network, mainly through the export of motor vehicles and components of foreign OEMs. But automobile industry in Belgium is facing a crisis. The chance of Belgium is seen in non-premium OEMs from China and India, which enter new foreign markets, especially with fuel-efficient cars. The market access of the Asian OEMs usually takes place on smaller automobile markets (e.g. Belgium, Finland) in combination with progressive price policies and high quality standards (Sandau and Wiegmann 2013). For the import of vehicles they need professionals of the European internal market as well as a trade network. With the international hub in

Zeebrugge and international OEMs, Belgium seemed to fulfill the necessary conditions as the "Gate to Europe".

In contrast, the rising automobile industries in the Slovak Republic and Slovenia results from the relocation of the automobile production and vendor parts production to Eastern European countries due to increasing labor cost in the west and south of Europe. In the Slovak Republic the three OEMs Volkswagen, Kia Motors and PSA Peugeot Citroen are responsible for one third of the Slovakian export volume (GTAI 2013). Slovakian 1st and 2nd tier suppliers export their automobile products to neighbor countries like Germany, Czech Republic, Hungary and Russia (GTAI 2013). Nearly the same in relation to production and export conditions could apply for Slovenia. Especially low trading barriers through geographical proximity of internal markets provide a win-win-situation for OEMs and nations productivity. Another reason results from the need of smaller countries to export products due to the small domestic sales market. The export orientation leads to deviations from the home market preferences and technological specifications. It has to considered, that the presence of export advantage in the named countries is still the result of relocation processes. Domestic suppliers profit from the export orientation and manufacture constraints of the foreign firms. In contrast, the example of the airbag shows the ability of respond to costumers needs by German OEMs and increasing exports (4th position). The global dominant innovation design of the airbag was a reaction to the greatest German automobile export market in the USA (Beise 2006). Relating to the personnel's safety, the necessity of reducing the frequency and seriousness of accidents by the low sensibility of using seat belts in the USA, German OEMs developed the airbag for premium vehicles.

## *5.5 Demand Advantage*

Considering the purchase price of a BEV, early adopters should have a higher willingness to pay for an electric vehicle than for a conventional propulsion system. In the PCA, this is included by private household expenditures for vehicles. The vehicle expenditures have a positive impact on the amount of charging stations. Further, the road infrastructure, represented by the roads total network in kilometers, plays a role in the development of an e-mobility lead market. Countries with a developed road infrastructure seem to be more automotive-oriented and afford a high-level mobility. These aspects facilitate the demonstration and visibility effect of diffusion of new automobile technologies. Additionally, countries with a demand advantage are characterized by a relatively high rate of inward direct foreign investments related to improved economic growth and advanced technology. This allows foreign companies to get access to resources and development opportunities of a domestic market. The domestic market could have an advanced adoption rate, which has a positive impact on the sale volumes of BEVs. The tested variables have a positive impact on each other. If the share of one variable increases, the other variables increase, too.

According to the regression coefficient, the USA own the demand advantage followed by Germany and China. In comparison to other countries, the household expenditures for vehicles are the highest in the USA. In the USA the share of vehicles per household and the average mileage per vehicle are higher than in other countries. The driving behavior of US households (high car usage) leads to higher total costs despite lower relative costs (e.g. gasoline costs). This could have a positive impact on the registered number of 69,106 between 2009 and 2012 (Statista 2013). The number of charging stations grows up from nearly 6,000 in 2012 to about 8,500 stations until now (U.S. Department of Energy 2013, 2014). The number of registered BEVs in Germany (7,491) and China (9,081) were even below those from France (15,309), Norway (10,005) and Japan (34,658) over the same period (Statista 2013). In Germany the number of charging stations grows from 2,200 (2012) to 4,400 in 2014 (AEE 2014).[6]

On the one hand, the US market is the greatest sales market for most innovation designs, which could be a sign for a large, isolated market (Beise 2006). On the other hand, the US government mainly introduces subsidies by creating purchase incentives for BEVs (e.g. purchase rebates, sales taxes and free installation of private charging stations) instead of market-protecting measures (U.S. Department of Energy 2013). It therefore seems more likely that China becomes a large, isolated market. It is known that Chinese product markets are often protected from foreign imports. Besides, the relative advantage of China is an effect of sales market size, high inward direct investments and the absolute roads total network.[7] The German government followed its objective of increasing the contribution of e-mobility through a holistic policy approach addressing specific and crossover topics of research and technology policy, energy policy, spatial and urban development and transport policy (BMVI 2014). It refers to requirements, such as transfer and promotion of the innovative capacity of the German economy. Despite relatively low variable values of the e-mobility indicators (BEV demand, number of charging stations), Germany profit of its advanced transport infrastructure and high inward direct investments.

## 5.6 Lead Market Potential in Summary

As mentioned above, lead markets only exist under multi-dimensional conditions. If a country is accounted for the lead market because of one advantage, the result could not be significant (Beise 2006). Testing the position through several ranking methods aggregates the values and verifies the results. Figure 2 shows the lead market position of the countries by the filters arithmetic mean, arithmetic mean

[6]There are no values available for China.

[7]Per capita values had to remove because of low communalities.

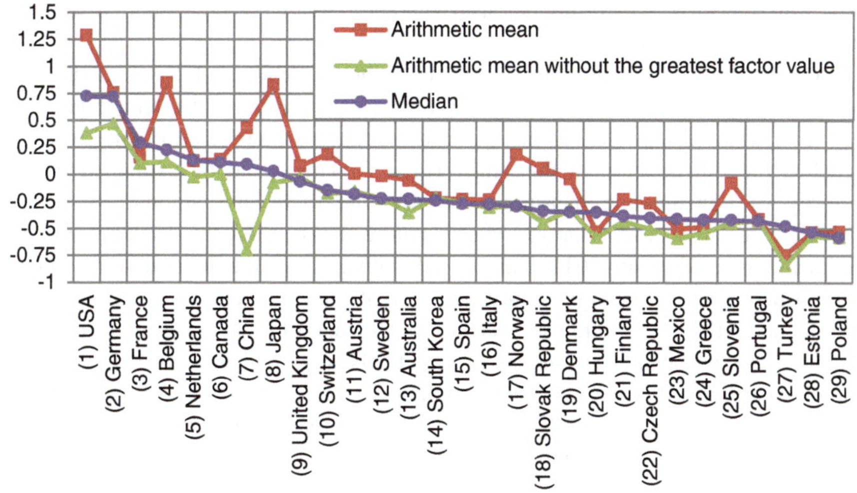


**Fig. 2** The lead market potential of countries derived by several ranking methods of the lead market factor regression coefficient (own calculation)

without the greatest factor value per country and median. The countries' order is based on the median.[8]

The USA have the highest lead market potenial. Especially the demand for e-mobility products accelarated in the last years, which was positively influenced by demand-pull legislations (tax credits, rebates and non-finanancial incentives) of the U.S. states governments. The political is completed by loans for OEMs for production and investments in modified production platforms and the evolution of hybrid concepts. Most of the electric vehicles and components are still US imports. The 2nd position is held by Germany followed by France, Belgium and the Netherlands. Germany's position results mainly from strong production and R&D activities in the global automobile industry. It has no obvious factor advantage. Generally, Germany is well advanced (e.g. transport infrastructure, income, export, high-tech products) and a favorite investment and joint venture partner in the automobile industry. But the lead market model counts for multi-factor advantages of supply and demand side activities. Hence, Germany offers the best e-mobility technology, but lose market shares due to the focus on e-mobility in premium cars (Roland Berger and fka 2014b). The adavantage of France is mainly a result of R&D activities as well as supply-side and demand-side incentives of the government. Especially Renault has competitive advantages with alternative engines through an early market entry. French OEMs are oriented towards customer needs and provide a basis for continuous developments. The lead market position of Belgium was already a theme of advantages in the price and export factor, which

[8]Median is the middle value of all measurement values, so that half of the values lie under the median and half over. The median is stable to outliers.

was not expected. Therefore the lead market potential is a results of ambient automobile market conditions.

At least the lead market potential of Japan should be focused. Since 2010, there are regularly lead market rankings of different consultants (RolandBerger, Mc Kinsey), where Japan secured the top lead market position for e-mobility, while it reached the 8th position in this study (Fig. 2). This could be explained by the priority of demand-side measures. Initially, the increasing amount of electric vehicle sales was mostly the result of a nature disaster. As the Tsunami and the Fukushima accident in 2011 had caused damages on the infrastructure, Japan established a stricter energy policy. As part of the emergency plan, BEVs from Nissan and Mitsubishi were used because of the lack of fuels in the affected zones. This led to new business strategies of the OEMs (e.g. vehicle-to-home-strategies), which were applied in the new energy policy (Bierau et al. 2014). Additionally, the purchase price of Japanese BEVs on the domestic market sunk of at least 40 % of the average price in European countries (RolandBerger and fka 2014a). This led to positive effects on the demand of e-mobility products by private households and the lead market position in other rankings. In this study, Japan has also characteristics of a large, isolated market, which cannot be distinguised at present. According to the lead market approach, the direction of technic evolution will be firstly decided on regional, domestic markets, which leads to an open window of opportunity for most OEMs to enter the market. Possibly, the evolution of e-mobility will be completed within a few years with the necessary compensated differences between the countries, a global dominant innovation design and political motivation.

## 6 Discussion and Conclusion

International and national organisations as well as MNCs can have substantial impact on the diffusion of innovations. The early market entry of a company leads to scale effects, falling prices and an increasing adoption rate. The development of a dominant innovation design to evolve to a lead market is a process of trial and error. This is even more likely, if the degree of market competition is highly combined with strong execution skills. In addition to first mover advantages, country-specific properties for e-mobility increase the chance that favoured innovations from a domestic market will diffuse abroad. But energy efficiency in the automobile industry is not necessarily widespread. Especially emerging markets with lower incomes and environmental protection legislations prefer powerful vehicles and accept lower efficiency standards. This puts some pressure to the MNC to meet the needs of different costumer groups. It can be expected that such countries and domestic manufactures follow other strategies than e-mobility.

The results show that the USA have an obvious advantage in demand and relative advantages across all lead market factors. This is mainly a result of demand-side incentives, a large sales market and low energy price levels. Compared to the USA, Germany reaches no direct factor advantage. The lead market potential of Germany mainly consists of supply-side instruments, well reputed automobile enterprises as well as foreign direct investments, export orientation and advanced R&D activities. We expected a larger lead market potential and advantages of Japan, which only exist in the market structure. It could be expected that Germany will become the leader in supply but apparently no lead market. The opportunities for the USA, France and Japan seem to be more advanced in multi-factor relations.

Considering the validity of data, further research is necessary to gain long-term metric data that is more closely connected to e-mobility. This was not feasible, when the study was carried out. Apparently, it is problematic to examine the lead market potential for different innovation designs in the same way. The size of countries and sales market (absolute vs. relative values) influences the result of the PCA. Additionally, the modeling approach of PCA is often criticized by the reduction of an amount of variables and the researcher influences in the process, whereby information gets lost. The lead market model has theoretical and conceptual weaknesses. The retrospective determination has often been criticized, especially according to appropriate policy measures for developing lead markets (Klodt 2011 amongst others). The applied model cannot achieve this. Further research is needed that requires substantial theoretical framework on the basic descriptions and statistical methods for measuring lead markets. Especially in the case of eco-innovation, which is often developed in globalized markets, the lead market approach proposed by Beise (2001) is not able to deliver adequate explanations (Köhler et al. 2014).

## References

Abernathy WJ (1978) The productivity dilemma: roadblock to innovation in the automobile industry. Johns Hopkins Press, Baltimore

ACEA (2013) Statistics. http://www.acea.be/statistics. Accessed 22 Oct 2013

AEE (2014) förderal erneuerbar. http://www.foederal-erneuerbar.de/startseite. Accessed 22 Sept 2014

Backhaus K, Erichson B, Plinke W, Weiber W (2006) Multivariate Analysemethoden: Eine anwendungsorientierte Einführung. Springer, Berlin

Bartlett C, Goshal S (1990) Managing innovation in the transnational corporation. In: Bartlett C, Doz Y, Hedlung G (eds) Managing the global firm. Routledge, London, pp 2154–2255

Beise M (2001) Lead markets. Country-specific success factors of the global diffusion of innovations. Physica, Heidelberg

Beise M (2006) Die Lead-Markt-Strategie. Das Geheimnis weltweit erfolgreicher Innovationen. Springer, Berlin

Beise M, Cleff T (2004) Assessing the lead market potential of countries for innovation projects. J Int Manag 10:453–477. http://dx.doi.org/10.1016/j.intman.2004.08.003

Beise M, Cleff T, Heneric O, Rammer C (2002) Lead Markt Deutschland: Zur Position Deutschlands als führender Absatzmarkt für Innovationen. Thematische Schwerpunktstudie im Rahmen der Berichterstattung zur technologischen Leistungsfähigkeit im Auftrag des bmb + f. Endbericht. http://hdl.handle.net/10419/39153. Accessed 30 Sept 2012

Bierau F, Müller B, Meyer G, Lidzba B, Schürmann M (2014) Trendbericht. Elektromobilität in Japan. Available via AHK Japan. http://www.japan.ahk.de/fileadmin/ahk_japan/Publikationen/

PDF/Trendbericht_Elektromobilitat_in_Japan_Februar_2014_Klein.pdf. Accessed 20 Jun 2014

BMVI (2011) Electric mobility—Germany as a lead market and lead provider. Available via BMVI. http://www.bmvi.de/SharedDocs/EN/Publikationen/electric-mobility-germany-as-a-lead-market-and-lead-provider.pdf?__blob=publicationFile. Accessed 10 June 2012

BMVI (2014) Verkehr und Mobilität. Modellregionen Elektromobilität. Available via BMVI. http://www.bmvi.de/SharedDocs/DE/Artikel/UI/modellregionen-elektromobilitaet.html?nn=36210. Accessed 05 June 2014

Cleff T, Rennings K (2011) Theoretical and empirical evidence of timing-to-market and lead market strategies for successful environmental innovation. Simon Fraser University, Vancouver, Canada. Available via SFU summit—institutional repository. http://summit.sfu.ca/system/files/iritems1/10921/Theoretical%20and%20Empirical%20Evidence%20of%20Timing-to-Market%20and%20Lead%20Market%20Strategies%20for%20Successful%20Environmental%20Innovation.pdf. Accessed 21 Sept 2013

Cleff T, Rennings K (2013) Do first mover advantages for producers of energy efficient appliances exist? The case of refrigerators. Available via ZEW Discussion Papers. http://ftp.zew.de/pub/zew-docs/dp/dp13054.pdf. Accessed 15 April 2013

Ganesh J, Kumar V (1996) Capturing the cross-national learning effect: an anlysis of an industrial technology diffusion. J Acad Market Sci 25(4):328–337. doi:10.1177/0092070396244004

Golder PN, Tellis GJ (1996) First to market, first to fail? Real causes of enduring market leadership. Sloan Manag Rev 37:65–75. http://www-rcf.usc.edu/~tellis/First.pdf. Accessed 6 June 2012

GTAI (2013) Trade. Internationale Märkte. http://www.gtai.de/GTAI/Navigation/DE/Trade/maerkte,did=951424.html. Accessed 10 June 2014

GTAI (2014) Branche kompakt - Kfz-Industrie und Kfz-Teile - Belgien, 2014. http://www.gtai.de/GTAI/Navigation/DE/Trade/maerkte,did=951424.html. Accessed 10 June 2014

IEA (2013) World energy outlook 2013. http://www.worldenergyoutlook.org/publications/weo-2013/. Accessed 10 Jan 2014

IRI (2013) The 2012 EU industrial R&D investment scoreboard. http://iri.jrc.ec.europa.eu/scoreboard12.html. Accessed 20 Oct 2013

Johansson JK (2009) Global marketing. Foreign entry, local marketing and global management. Mcgraw-Hill/Irwin, New York

Klodt H (2011) Zum Konzept der Leitmärkte – Ein Einführung. In: BMBF (ed) Wettbewerbsfähiger durch Leitmarktstrategie? Available via TU Berlin http://www.inno.tu-berlin.de/fileadmin/a38335100/PDF_Dateien/Dokumentation_BMBF_Workshop_Leitmarktstrategie.pdf. Accessed 29 Dec 2013

Köhler J, Walz R, Marscheder-Weidemann F, Thedieck B (2014) Lead markets in 2nd generation biofuels for aviation: a comparison of Germany, Brazil and the USA. Environ Innov Soc Trans 10:59–76. doi:10.1016/j.eist.2013.10.003

Lieberman MB, Montgomery DB (1987) First-mover-advantages. Strateg Manag J 9:41–58

Lieberman MB, Montgomery DB (1998) First-mover (dis)advantages: retrospective and link with the resourced-based view. Strateg Manag J 19:1111–1125

Meyer-Krahmer F, Reger G (1997) Konsequenzen veränderter industrieller F&E-Strategien für die nationale Forschungs- und Technologiepolitik. In: Gerybadze A, Meyer-Krahmer F, Reger G (eds) Globales Management von Forschung und Innovation. Schäffer-Poeschel, Stuttgart, pp 196–215

Murmann JP, Frenken K (2006) Towards a systematic framework for research on dominant designs, technological innovations, and industrial change. Res Policy 35:925–952. doi:10.1016/j.respol.2006.04.011

Nokia (2014) Our company. Our history. http://company.nokia.com/en/about-us/our-company/our-story. Accessed 18 Feb 2014

OECD (2013) OECD data lab. http://www.oecd.org/statistics/. Accessed 21 Oct 2013

OICA (2013) Production statistics. http://www.oica.net/category/production-statistics/. Accessed 22 Oct 2013

Ohmae K (1995) The end of the nation state: the rise of the regional economies. Free Press, New York

Peters A, Dütschke E (2014) How do consumers perceive electric vehicles? A comparison of german consumer groups. J Environ Policy Plann. http://dx.doi.org/10.1080/1523908X.2013.879037

Quitzow R (2013) Towards an integrated approach to promoting environmental innovation and national competiveness. Innov Dev 3(2):277–296

REN21 (2012) Renewables 2012. Global status report. http://www.ren21.net/Portals/0/documents/Resources/GSR2012_low%20res_FINAL.pdf. Accessed 01 Feb 2013

Rennings K, Beise M (2003) Lead markets of environmental innovations: a framework for innovation and environmental economics. ZEW discussion papers, no 03-01. http://hdl.handle.net/10419/24806. Accessed 20 Nov 2012

Rogers EM (2003) Diffusion of innovations. Free Press, New York

Roland Berger and fka (2014a) E-Mobilität: Japan baut Industrieführerschaft und Absatzzahlen aus – deutliche Preissenkungen und breite Lade-Infrastruktur sind Wachstumstreibe. http://www.rolandberger.de/expertise/branchenexpertise/green_business/Index_Elektromobilitaet_zeigt_Japans_Fuehrerschaft. Accessed 13 June 2014

Roland Berger and fka (2014b) E-Mobilität: Deutschland erobert Technologieführerschaft zurück – Absatzzahlen liegen aber weiterhin nur im Mittelfeld. http://www.fka.de/pdf/presse/2014-09-19-14fka0004.pdf. Accessed 25 June 2014

Rosenberg N (1982) Inside the black box: technology and economics. Cambridge University Press, Cambridge

Rosenbloom RS, Christensen CM (1995) Explaining the attackers advantages: technological paradigms, organizational dynamics, and the value network. Res Policy 24:233–257

Sandau T, Wiegmann J (2013) Die Wettbewerbsfähigkeit der europäischen Automobilindustrie auf den Märkten von morgen. In: Proff H, Pascha W, Schönharting J, Schramm D (eds) Schritte in die zukünftige Mobilität. Technische und betriebswirtschaftliche Aspekte. Springer, Wiesbaden, pp 509–530. doi:10.1007/978-3-8349-4308-8_32

Statista (2013) Elektromobilität – Statista Dossier 2012. http://de.statista.com/statistik/studie/id/6547/dokument/elektromobilitaet-statista-dossier/. Accessed 10 Oct 2013

Suarez F, Lanzolla G (2005) The half-truth of first-mover advantage. Harvard Bus Rev 83(4):121–128

US Department of Energy (2013) AFDC alternative fuel data center. http://www.energy.gov/maps/alternative-fueling-station-locator. Accessed 10 Oct 2013

US Department of Energy (2014) AFDC alternative fuel data center. http://www.afdc.energy.gov/locator/stations/?fuel=ELEC. Accessed 01 Sept 2014

UNcomtrade (2013) United Nations commodity trade statistics database. http://comtrade.un.org/db/mr/daCommodities.aspx. Accessed 01 Oct 2013

Worldbank (2012) World bank open data. http://data.worldbank.org/. Accessed 01 Feb 2013

Worldbank (2013) World bank open data. http://data.worldbank.org/. Accessed 21 Oct 2013

## Author Biography

**Noreen E. Werner** has been a research associate at the Centre of Regional and Innovation Economics at the University of Bremen. She studied Geography at the Friedrich-Schiller-University of Jena and urban and regional development at the University of Bremen. Her research interests are innovation management, lead markets, mobility pattern and regional development with a view on e-mobility and political influences.

